SURVIVING THE WHITEBOARD INTERVIEW

A DEVELOPER'S GUIDE TO USING SOFT SKILLS TO GET HIRED

William Gant

Apress®

Surviving the Whiteboard Interview: A Developer's Guide to Using Soft Skills to Get Hired

William Gant
Nashville, TN, USA

ISBN-13 (pbk): 978-1-4842-5006-8   ISBN-13 (electronic): 978-1-4842-5007-5
https://doi.org/10.1007/978-1-4842-5007-5

Copyright © 2019 by William Gant

Managing Director, Apress Media LLC: Welmoed Spahr
Acquisitions Editor: Shiva Ramachandran
Development Editor: Rita Fernando
Coordinating Editor: Rita Fernando

Cover designed by eStudioCalamar

Distributed to the book trade worldwide by Springer Science+Business Media New York, 233 Spring Street, 6th Floor, New York, NY 10013. Phone 1-800-SPRINGER, fax (201) 348-4505, e-mail orders-ny@springer-sbm.com, or visit www.springeronline.com. Apress Media, LLC is a California LLC and the sole member (owner) is Springer Science + Business Media Finance Inc (SSBM Finance Inc). SSBM Finance Inc is a **Delaware** corporation.

For information on translations, please e-mail rights@apress.com, or visit www.apress.com/rights-permissions.

Apress titles may be purchased in bulk for academic, corporate, or promotional use. eBook versions and licenses are also available for most titles. For more information, reference our Print and eBook Bulk Sales web page at www.apress.com/bulk-sales.

Any source code or other supplementary material referenced by the author in this book is available to readers on GitHub via the book's product page, located at www.apress.com/9781484250068. For more detailed information, please visit www.apress.com/source-code.

Printed on acid-free paper

This book is dedicated to all those who stretch the definition of "possible" and make it not only achievable but who show the way to make it the "everyday" for others. I especially dedicate it to my friends Doc, Rob, James, David, and John, without whom my world would be smaller and the possibilities within it dimmer.

This book is also dedicated to my wife, who has exhibited a remarkable level of calm for one subjected first to a budding programmer, then to a budding podcaster, and finally to a budding self-marketing writer and public speaker. That's not a thing to be underestimated.

Contents

About the Author

William Gant is a software developer who has been working in the industry since 1998. He has worked with dozens of different programming languages in a wide variety of programming environments, both as an employee and as a consultant. In addition, he has owned his own business, been heavily involved in several startups, and is currently half of the Complete Developer Podcast, among numerous other things in the development space.

Acknowledgments

Many thanks to Erin Orstrom. A question of hers during a meetup group long ago started the chain of events that culminated in this book, and she did the technical review of the code in Chapter 4.

Introduction

You might be wondering why I am qualified to write this book. You'd be right to do so. Lots of stuff is available about getting through the development interview process, some by experienced developers and others that, frankly, don't have a wide range of experience to draw from.

I, on the other hand, do have a fair bit of experience in this space. November 2018 marked a milestone for me, because that month will be the 20th anniversary of the first time I got paid for writing code, as well as the anniversary of the first time I went through the process of a development interview. That was my freshman year of college and happened after four and a half years of learning to code on my own. I later followed that up with completing my bachelor's at Lipscomb University, hitting the workforce and interviewing for full-time positions in late 2002. Since then, I've had 17 software development jobs, many of which were short-term contracts, along with numerous side projects, most of which required one or more interviews and at least half of which required getting through a whiteboard interview.

I've not only survived numerous software development interviews from the interviewee side of the table, but I have conducted interviews as both a manager and as a lead developer. Having led and managed developers, I've learned a lot about what a manager is looking for when conducting an interview. Knowing this has given me a lot of insight into some best practices for developers who are trying to get hired.

While not a qualification for writing this guide, I'd also like to point out here that I am one of the guys on the Complete Developer Podcast (www. CompleteDeveloperPodcast.com). As of the writing of this book, the podcast has been going for more than 3 years and has a lot of stuff that might appeal to you if you are reading this guide. In addition, I am also one of the founders of Developer Launchpad (www.DeveloperLaunchpad.com), which is a meetup group focused around helping developers get ready for developer interviews by giving them practice coding exercises and real training on dealing with whiteboard interviews, so that their first experience with it isn't in an interview where the stakes are higher. I also frequently speak on the topic and have presented a talk on this topic at numerous times.

I am sharing these tactics with you in this book for a lot of reasons. For one, whiteboard interviews are unnecessarily challenging and difficult, especially for new developers who don't know the tricks and tactics to get through it. I also personally detest whiteboard interviews, as I don't think they are a particularly good way to evaluate someone's skill as a developer. I hope that this guide will not only help you get through this process in a way that makes it work well for you, but that the hacks that this book contains will circulate widely enough to eventually make it a less popular interviewing method.

Developer hiring is fundamentally broken as currently constructed, and if enough people can learn to use the broken system to their advantage, we just might rid ourselves of it. In that sense, this serves as a bit of graffiti on the wall that keeps qualified people out of development when they shouldn't be. You might say that this is a bit of a personal crusade and you'd be right.

Why Software Development Interviews Are Hard

In this chapter, we'll start out by examining why so many of us find interviews, especially in software development, to be such a painful and difficult experience. We'll start out discussing physiological responses and how to manage them. Next, we'll discuss some strategies for turning the psychological responses into something that is actually useful. Finally, we'll examine several ways to prepare for interviews that we will discuss in further detail later in the book.

I'll start this out with the same question that I always ask when I'm giving a presentation on surviving whiteboard interviews. That question is: "How many of you find interviews frightening?" When I ask a classroom full of aspiring software developers this question, nearly all of them raise their hands. Do you know why this makes you nervous? It's because you are rational. That's all. It's not a personality defect. It's just evidence that you are a normal human being.

© William Gant 2019
W. Gant, *Surviving the Whiteboard Interview*,
https://doi.org/10.1007/978-1-4842-5007-5_1

There are a lot of very valid reasons to find an interview nerve-wracking. You can fail and if you do, you get to go through the entire awful process somewhere else. Not only do you have the risk of failure but you probably feel like the interview process is unpredictable. Going in, you have no idea how long the interview will take, who is going to be asking you questions, and what the format will be. It can create a sense of helplessness, as if the interview is something being inflicted upon you rather than something you chose to subject yourself to.

A whiteboard interview is often the worst case. It combines the worst aspects of a standard interview, public speaking, a quiz show with a hostile host, and a dinner party full of strangers who are silently (or not so silently) judging you. Very little of it probably feels much like your reasons for getting into software development, and you start to wonder whether the experience is a valid reason to get out of it. I personally have seen multiple brilliant developers who are the best that I know, who nevertheless completely bombed in front of a whiteboard. Worse still, I've seen terrible developers who have had great success in the interview process, only to end up getting fired 6 months later because they couldn't do their jobs.

Let's face it. Not very many people who decide to make a life out of software development really relish the thought of being put on the spot in front of a group of strangers. If we wanted to do that, it's pretty likely that we would have made an entirely different career choice in the first place, as development tends to be a bit more isolated than many other career paths. You are essentially being asked to make a sale, and you probably aren't very comfortable with doing so. Being the center of others' attention is made even worse by the fact that they are usually complete strangers that you just met less than an hour before. In short, it's the worst aspects of petty socializing, cold sales, and public speaking, all rolled into one.

Even worse though is the thought process you are likely to go through in this situation. If this interview doesn't go well, what happens then? Your ability to pay your bills may well be on the line. Even if you still have income while this is going on, the downside of failure in this is that you either have to give up or you get the pleasure of going through it again later, with a different problem and a different set of random strangers. Even better, because you don't necessarily know what kind of questions you'll be grilled on, you can't even reassure yourself that next time will be better, because it could very easily be worse. If you are new to coding and this is your first whiteboarding experience, it's even worse, because it's all brand new. It's this cycle of thoughts that we must break.

Physiological Responses

As a result of all this going through your mind, your body starts reacting as well. Human beings have a fairly limited range of physiological responses to nervousness, and those responses were developed over millennia for dealing with things like predators. The response is thus frequently a bit over the top for something that seems as benign as a job interview, but the way that the experience feels will still be the same. This set of physiological responses are the reason why a single misstep so often results in a more serious failure, as nervousness feeds into mistakes, which produce more nervousness. The cycle is liable to repeat until it is stopped.

A few things happen when you get really nervous, none of which are particularly helpful when you are trying to show someone how you can bring value to their team.

- Your heart rate will increase and you will begin sweating, often to excess.

- You will probably start speaking faster, and being less cautious about what you say.

- Should you have to focus during this experience (and you'd better), you are also far more likely to make mistakes.

In short, your built-in biological response to being nervous is more than capable of sabotaging an interview for you. It can be really unpleasant. This set of responses is supposed to keep you from getting eaten by a lion, and it makes you unable to pay your rent instead. In short, human beings evolved responses to stimuli that were appropriate when we were hunter-gatherers, but that have ceased to be helpful in the sort of modern, urban environments where most interviews take place.

Natural reactions to stimuli are very difficult to remove. Because they happen before you can even think about them, there isn't much you can do until after the fact. While mitigating these reactions is a useful and necessary tactic, the best way to keep things under control is frequent exposure to the stimulus. However, it is difficult to experience enough interviews in a short time to completely get rid of the very natural fear that they create. Experience (and the lack of fear that comes with it) will come in time, but that doesn't get rid of the immediate problem.

Instead, we're going to have to work with the reactions we have and try to turn them to our advantage.

It's In Your Head

Besides being in an interview, there is another situation where many of the same physiological responses occur, and that situation has nothing to do with avoiding being eaten. Instead, it is one of the best feelings in the world. When you are feeling joy and exhilaration, your heart rate will also increase, you will probably sweat more, speak faster, and even make more mistakes, yet most of us don't walk around in fear of being happy and excited.

Whether it is getting on a roller coaster, skydiving, driving fast, or playing a sport that you love, a feeling of exhilaration will produce very similar responses to fear, yet they won't feel terrible. In fact, you're very likely to have a heightened sense of focus and to thoroughly enjoy yourself. Wouldn't it be great if you could feel the same way about a nasty programming whiteboard interview, and even look forward to the challenge?

The good news is that you largely can do that, because the difference between the two isn't physiological; it's psychological. At the end of the day, getting into the right mental state will help you succeed both on whiteboard challenges and in other challenges in your life. It is not only possible to do this, but to make that sweaty, rapid heartbeat, sour stomach, panic-inducing feeling that you get when being interviewed into a feeling of excitement at the challenge. And it's easier than you might think.

Do you remember the bit in *Star Wars: The Empire Strikes Back* (1980), where Yoda sends Luke into the cave, telling him that the only thing in there is what he takes with him? And then Luke runs into Vader and makes a mess of things? While sounding a lot like the sort of pop psychology stuff that we are all more than a little tired of, it does represent a statement of profound truth. You are going to encounter what you bring into the interview. If you go in scared to death and concerned about what happens when you fail, you drastically improve the odds of that failure. If, on the other hand, you go in with quiet confidence and well-practiced skill, your odds of success are much higher. That's just the way it goes.

There are three main differences between something that induces a feeling of panic and something that produces a feeling of exhilaration.

- You have to know you are "safe." In other words, you need to have the assurance that you won't be harmed by the experience.

- You have to be "happily" anticipating the experience. In other words, there has to be something positive to look forward to for enduring the experience.

- You have to feel some degree of control over the experience.

All of these characteristics are manageable, if you prepare in advance. Any good method of interview preparation must not only deal with the mechanics of interviewing but must also mitigate the psychological and physiological difficulties one will encounter while doing so. In addition, any preparation should provide a real sense of being in control of the process.

Get Your Stuff Together

Because you understand that most of the difficulty comes from your response to the interview, you know that your odds of success are going to increase if you are prepared in advance. As we work our way through this book together, we're going to put a plan of action together to make sure you are as well prepared as possible for the "exciting adventure" of going through a developer whiteboard interview. While I can't promise that you will ever truly "enjoy" interviewing, if you prepare properly, I can promise that it will no longer inspire fear. Toward this end, the rest of the book will explain how to prepare for an interview using the following tools and tactics:

- Practical drills to help you practice for the actual interview (Chapter 4).

- Tricks you can (and should) use to get through the challenging bits of interacting with strangers during the process (Chapter 9).

- Some history that will help you understand why software interviewing is broken, how we got here, and how you can use this information to your advantage (Chapter 2).

- What the interviewer is looking for and what little things you can change about your approach to be subtly better than the other people going through the process (Chapter 11).

- How first impressions impact the direction of the interview and how to make a better one (Chapters 5–9).

- How to control the direction of the interview by having a good resume (Chapter 5).

- What you should be learning about the employer when interviewing. After all, a great interview with a bad company is still a bad interview (Chapter 10).

- Some things to watch for when interviewing. Part of being safe is recognizing danger signs, after all (Chapters 10–11).

You don't want an interview to be the first time you experience a whiteboard problem, coding challenge, or group interview. Novelty makes nervousness much worse. Instead, you are far better off doing practice interviews and drills until the actual interview isn't bad at all. Provided you do well, a "boring" interview is not a bad thing at all.

Interacting effectively with people you don't know is also a challenge. Further, it's one that many people in tech find especially difficult. Interpersonal interactions with people you know are usually easier than those with complete strangers. Consequently, we will be going over some tricks you can use to quickly build rapport with the interviewer. This will help make you stand out from other interviewees.

If you understand the history of technology interviews, you'll be better equipped to handle the various ways that companies try to innovate on the process. While many companies do not have very interesting interview processes, you'll occasionally come across one that tries something new and different. However, if you understand what interviewers are trying to learn along with the history of how they try to learn it, you'll be better prepared to work your way around the more "interesting" approaches to interviewing that you'll eventually encounter.

You can also exert a fair amount of control over the interview process by being careful about how you present yourself. The first few seconds of an interaction provide a ton of information to the interviewer, and bad impressions are hard to overcome. Not only are the first few minutes of an interview critical, but so is a well-constructed resume. Both can either be assets to you or can harm you in a way that causes you not to get the job.

In addition to preparing effectively, you also need to control the frame of the questions in the interview by having questions of your own. You do not want to come to an interview begging for a job; instead, you want to come to an interview in an effort to find a mutually beneficial arrangement for yourself and the employer. You need to know what you are getting into, as you don't want to succeed at an interview with a bad company (trust me). This is also a good time to gather information that will be useful for making a positive impact on your first day.

Summary

In this chapter, we discussed why your built-in physiological and psychological responses to stress can make interviews especially difficult. Next, we discussed why these responses can't easily be overcome and what to do instead of trying to overcome them. We also discussed a strategy for dealing with these responses that can make them a lot easier to bear. Finally, we went through a brief overview of the sorts of things you should be doing to prepare for an interview, along with a brief discussion of how to handle the interview itself.

Why Software Development Hiring Is Broken

Hiring in software development is horribly, perhaps irretrievably, broken. As both a software development manager and a software developer, I've seen the process stop good developers from getting jobs while incompetent numbskulls who shouldn't be anywhere near production get hired over and over again. The latter do so only to get fired a few months later when they can't deliver results. Usually the rest of us get to clean up their results for months or years after they are gone. It's really dumb that most places use a broken process, but that's the reality we have to deal with.

In this chapter, we're going to discuss the reasons why it is difficult to evaluate software developers, how companies have traditionally tried to handle it, and how various approaches have failed. Understanding why things are messed up is very helpful in dealing with the problem in a constructive way, even if you can't fix the problem itself. Software development hiring managers have been trying to find a good way to evaluate developers for decades, and have yet to come up with anything that both works well and is trivial to implement. We'll discuss how the industry first used quizzes, live coding challenges, and various other approaches to evaluate developers. Then we will discuss how the industry settled on using whiteboard interviews as a common way to evaluate developers.

© William Gant 2019
W. Gant, *Surviving the Whiteboard Interview*,
https://doi.org/10.1007/978-1-4842-5007-5_2

Difficult to Evaluate

The biggest problem of the software development process is that it is difficult to evaluate candidates for a variety of reasons, especially when the market is tight. If you hire someone unqualified, they can (and will) easily cost you thousands or tens of thousands of dollars before you get to go through the process again. All that is assuming that they don't take the company down with them, putting you in a position where you are interviewing from the other side of the table. On the other hand, if you do manage to get the resume of a good developer across your desk and the market is full of jobs, you may not get a lot of time to decide whether you want to hire this person or not, before they get snapped up by another company in town. Hiring managers are in a very difficult position and are thus forced to make a lot of decisions quickly, even if they come to regret those decisions later.

Worse still, hiring managers don't have a lot of time. If a healthy company is hiring, it's because they either have more work than they can do or they are anticipating having more work than they can do. Neither situation gives the hiring manager a lot of time to carefully consider candidates. This tends to mean that they will develop shortcuts to cut down on how much time they spend on the hiring process. This is especially true of software development managers who are still writing code themselves. Many software development managers don't enjoy managing people and will go to a lot of effort to limit their time doing so. As a result, you are very likely to be interviewed by someone who is in a hurry, doesn't particularly want to be there, and just wants to be done with the whole thing so they can get back to the rest of the work on their plate. This state of affairs can be a huge disadvantage to the unprepared, but can be wonderfully helpful to those who have done the work up front to make it an advantage.

Software development is unusual, especially for such a high-paying profession, in that there are a ton of ways that someone can get into the industry, with varying degrees of success. Not all software developers go to school for it; in fact, on the majority of teams I've worked with, most developers haven't had a computer science degree. And it's not like such developers lack skill—my friend Jeremiah doesn't have such a degree, and I know he can code circles around me. The difference is made with diligent practice, work ethic, and experience. The degree can help, but is not a sufficient measurement on its own.

And there's another problem too. Not everyone that goes to school is good. In fact, I've worked with a number of people who have gone as far as getting a PhD in computer science who, being kind, can't code their way out of a wet paper bag. You can't use a degree as proof of skill in software development, because it isn't one. Many companies try to use a computer science degree to evaluate people and often end up passing up on candidates that would have been an excellent fit for them. I've seen good candidates ignored and bad candidates hired, solely on the basis of a degree.

Yet another issue occurs due to this lack of focus on credentials and high pay. These conditions often attract people who wouldn't be pursuing this field if it weren't for the pay. As both a development manager and a developer, I've interacted with dozens of such people, including a very scruffy 20-year-old who was loudly certain he could get a job paying $70,000 a year because he "knew" how to build web applications. As it turned out, what he had done wasn't any more complicated than what you'd expect out of someone who had just had an introduction to HTML and had been halfway asleep while getting it. I guarantee you that any experienced hiring manager can tell dozens of similar stories. Not only do they have to evaluate the top people, but they have to filter this sort out before they waste too much time.

Busted Problems, Busted Solutions

Since the 1990s, when learning how to code became far easier and accessible, the industry has tried a variety of hacks in a vain attempt to quickly evaluate developers and get around these problems. All of these attempts had issues that meant that they didn't live up to their promise. However, by examining the shattered wreckage of these bad ideas, we can learn a lot about what kinds of things hiring managers are trying to determine in the space of a short interview.

The Development Quiz

When I started trying to get paid for writing code, most development shops were trying to solve this problem with quizzes. Because it was the 1990s and early 2000s, the Internet played far less of a role in these processes initially, since it was slow and hard to search, and web applications were very limited. Essentially, the idea was that you would be presented with a printout with a series of questions about your chosen platform, and your answers to those questions would essentially be used to evaluate your suitability for a job. Many places developed their own questions in house, based upon the needs of the business. There was a short period of time where this worked really well, because the hiring manager's questions for the developer came right out of their own experience and reflected the kind of problems that one might encounter on the job.

However, it wasn't going to last. As the dotcom bubble approached and passed by, two things spelled the end of the in-house development quiz. The first was the wide availability of decent Internet search engines (stuff like Dogpile and Lycos) which allowed hiring managers to simply search for quizzes to give, rather than carefully formulating their own. However, the same technological change also allowed savvy developers to look up the same quizzes and prepare ahead of time. Between about 2002 and 2008, I bet I saw the same quiz in at least a dozen interviews on C#. As a result of these changes, the value of the in-house quiz became lower by the day, becoming roughly useless in a matter of months. This resulted in companies abandoning the practice within a few years.

Such a scenario is quite bad for developers, because an environment where everyone has the same answers to the same questions is an environment where you can't differentiate yourself as easily. You may see a few companies still using this, but it isn't as common since people figured out how to hack it. If you do see this in use, you can learn a lot by comparing the questions they ask you to those that are commonly available on the Internet. If they've carefully formulated questions that are specific to their needs, that's a good sign. If the list of questions is one that you've seen before, then they aren't being as careful as they probably should be. However, if you can do well on this part of the interview, it will still help you.

Development quizzes offer a few advantages to the interviewer. First, it is an approach that scales well, so they can give the same quiz to a large number of people without greatly increasing the amount of effort they are expending. Secondly, this approach allows nondevelopment managers to use the quiz as an initial screening mechanism. You'll often see this when you have an initial "phone screen" with Human Resources personnel—the quiz is intended to filter out a large percentage of candidates so that the hiring manager can choose among the best. Finally, such a quiz is low maintenance. Whether a company develops their own or gets one from the Internet, they don't really have to do much to maintain the quiz. It will generally work as well in a year as it does today.

Because of these advantages, development quizzes are still something that many hiring managers have in their arsenal. However, it was soon to be superseded by other approaches that were a little harder to game.

Coding Challenges

Another approach that became common later on was the coding challenge. While a few places let you take the challenge home and bring it back in, most actually required you to sit at a computer on-site to complete the challenge. The take-home challenges tended to be more difficult, while the in-house ones were simpler. There were a lot of upsides and downsides to both approaches.

Take-home challenges offered a great way for companies to evaluate a lot of developers concurrently. Instead of tying up the team while someone coded, this approach allowed companies to screen developers in a way that scaled pretty well. If the developer didn't complete the challenge, they were no longer in consideration. However, this approach had a lot of problems. First, and most obvious, it made cheating easy, whether such cheating was done through one's own circle of friends or via the Internet. Second, many developers balked at these challenges, feeling that they were essentially being asked to do work for free. Finally, while the practice was very scalable for hiring managers who wanted to test out a lot of developers, it didn't work very well for job seekers who wanted to apply at many positions. Once a developer has completed a take-home challenge or two without being hired, they are more likely to avoid them in the future.

A few companies tried to do coding challenges in house, but there were problems with this approach as well. First, the challenges couldn't be very large, because it actually wasted the company's time (instead of "just" the developer's time). Second, and perhaps most important, it also meant that a random stranger off the street was going to be using a computer inside the corporate network. This either required a lot of preparation up front or constituted a risk that most companies weren't willing to take once they thought about it a little.

Live coding with people watching you is almost as bad as a whiteboard interview for many of the same reasons, but it probably would be an improvement. Nevertheless, this approach is not something you are going to see much anymore. That said, if you do see a company using this approach, it's potentially a very good sign, as it means that their pool of applicants is smaller. Competing against fewer people is often easier, so long as you can compete effectively.

Whiteboard Interviews

As a result of all this experimentation, the industry ended up with the much-feared method of programmer evaluation that is a major subject of this book. This process came about for a variety of reasons that seemed like a good idea at the time.

First, interviewers are pressed for time. It's uncommon for companies to hire when they don't already have a need. As a result, the people involved in the interview process probably already have more work to do than they can expect to complete. The whiteboard interview theoretically offers a pretty reasonable way for them to evaluate developers without spending a ton of time, giving them the opportunity to cheat, or giving them access to computers inside the company network.

It's also very common for interviews to include an interview with the team to "assess cultural fit," which is an HR euphemism for the process of figuring out whether you and the team can stand each other. These usually happen in conference rooms because of seating constraints. This process will be discussed shortly, but the basic idea is to seat everyone together and discuss why you might be a good candidate for the job. Afterward, the team will be consulted to see what kind of impression you made.

As a result of these things, along with the near certain presence of whiteboards in conference rooms, the whiteboarding process was a natural fit for the way many organizations were already doing things. Add to that the ability to quickly find sample whiteboard problems on the Internet, and it totally makes sense why this happened. Whiteboard interviews have thus become very common, and are often the first choice, especially for stressed-out hiring managers.

All this history is important, not because you are going to be given a test on how interviewing practices have evolved but because the history itself points to the kinds of problems that interviewers are trying to solve using the whiteboard. If you know those things, it's easier to make appropriate decisions about what to do based on what the interviewer is actually looking for.

A whiteboard interview also tells you a few things about the company. First, they have a pretty standard approach to software development hiring, and they probably (mostly) interview average developers. Depending on your career situation and goals, this can either be great news or can convince you to look elsewhere (we'll discuss this in a later chapter). While it may not sound so great for a company to just be interviewing "average" developers, there are a lot of reasons why you might want to work at such a company.

Evaluating Soft Skills

A few things also make coding skills even more difficult to evaluate than normal. While you as an interviewee may not be concerned about these things, you might want to start thinking about how to use them to your advantage. In this section, we'll discuss things that make interviews even harder for the interviewer. Later, we'll discuss how to turn them to your advantage as an interviewee.

First, interviews and live coding exercises done remotely are often more difficult to evaluate. You lose a lot of useful information when the other party in an interview is remote. Besides all the technological problems that will invariably occur during such an interview, you also tend to miss most of the body language that the other side is using. Body language is a significant portion of our communication, and there is a greater chance of miscommunication when it is absent. At best, miscommunication makes interviews more frustrating and longer. At worst, it can completely derail the entire interview. These things are true for both parties in the interview.

Second, interviews involving a language or major cultural barrier are more difficult. This doesn't mean that you should avoid interviewing somewhere where the interviewers speak a different language or come from a different culture. Rather, it means that communications will be more difficult. Besides miscommunication from language barriers, cultural barriers can also make things more challenging. Even cultural differences within the same country can represent large differences in manners, ways of communication, and even ways of expressing yourself. These problems easily result in miscommunication, as well as being possible sources of bias.

Third, interviews conducted by nontechnical personnel create a lot of challenges. Besides simply not knowing enough about the technology being used, nontechnical interviewers will often fixate on the things that they do know, even if those things have nothing to do with the job in question. While this doesn't necessarily work to your advantage, it does mean that the interviewer may not be able to answer many of your questions. This can make it more difficult to decide if the job is worthwhile.

Finally, if the interviewers don't agree on what constitutes a good candidate, it gets much more challenging to be perceived as one. This can happen because of a lack of clear organizational structure, a "hybrid" job, or because of internal politics, but it will make the interview more difficult no matter the reason. You may find that some of your answers please one party and irritate the other. It's a difficult situation, made more difficult because you don't understand the situation that caused it.

Characteristics of Good Evaluations

Many companies are also terrible at evaluating soft skills. While your ability to work well with people is often given less emphasis for technical jobs, it is still critical to the success of any project in which you are involved. There are a lot of reasons that companies have difficulty with evaluating soft skills.

The first (and simplest) reason that companies aren't good judges of soft skills is that at least some percentage of job candidates are dishonest. Developers command high salaries, and this money provides incentives for less-than-scrupulous people to act deceptively in interviews. Whether they are lying about previous situations, providing an inaccurate view of how they would really handle a job situation, or simply relying on flattery to get through the interview, the results are often the same. Six months after hiring such an individual, the company is often left trying to figure out how to fire them without getting sued. Even worse, such an individual can cause so many interpersonal problems in an office that the best staff members leave.

Secondly, interviews are usually short by design. Neither the interviewer nor the interviewee can afford to have their time wasted. It takes a while to get a good view of someone's personality, and such extended conversations are difficult in an interview. Many people with truly bad soft skills and interpersonal habits can come across as charming, fun people when you first meet them, and only become an obvious problem later when things start falling apart.

Third, interviewers often can't get a complete picture of a person's personality even by contacting references. It's especially difficult to find out about character flaws from references, either because they don't know or refuse to tell you. There is a legal risk to disclosing negative information that keeps someone from getting a job. While a lawsuit is not particularly likely, many companies are justifiably paranoid about giving any negative information about anyone when a stranger calls for a reference. It's fairly common practice, especially with larger companies, to only confirm that you worked for them and when—they won't tell anything else at all.

Finally, work habits are often very tricky to evaluate. Take, for instance, an employee that is frequently late to work. Are they late because they are inconsiderate and hate their job? Do they have a short-term family situation that is making it impossible to arrive on time? Are they just disorganized? The first case might get better if they were in a more satisfying job, while the second might get better on its own. The third situation might get better with time and maturity, or it might not... An employer might be more or less willing to work with someone in each of these situations, but they may not even know that punctuality is a problem for weeks or months after hiring someone.

While soft skills are critical for being a successful software developer, they are one of the most difficult and risky things for an employer to evaluate. The damage caused by employees with poor (or nonexistent) soft skills can last for years after those people are gone. In a later section, we will discuss some things you can do to make sure that you make a good impression.

Other Methods of Evaluation

While whiteboards are almost certainly the most common way to evaluate developer skills, there are a few other approaches that are better. Typically, these approaches attempt to combine the best of earlier approaches with good techniques for determining culture fit. You won't see these often, but they frequently are a sign of a forward-looking company.

First, a good evaluation method will show your problem-solving skills on a real problem. This means that it will show your design skills, your debugging skills, and your ability to deal with ambiguous (or even changing) requirements. Such an approach should show off your ability to ask relevant questions and show how quickly you can understand the underlying business requirements. This shows that you can be a useful team member, who doesn't simply lock themselves in an office and write code.

Second, a good evaluation method will show how you interact with a team. It should incorporate giving and receiving feedback. It should show your reaction to corrections and suggestions. It should also show how you interact with team members and whether you can do so in a healthy manner. This shows that you won't upset the existing team dynamics (if healthy) or, at the least, won't make them worse (if unhealthy). It also allows you to evaluate whether you can stand to work with these people.

Third, it should show that you can actually do the work. While many people think that the purpose of an interview is determining whether you can do the work, it is not the only consideration. However, you do need to show that you can actually get work done. A good evaluation should definitively prove that, without it being a matter of guesswork.

Fourth, a good evaluation process should give you an idea of what it is really like to work at a company. While you can get hints of that in an interview, the day-to-day grind of many jobs is something you have to experience firsthand to understand. Many companies have processes or just general ways of working that make them unpleasant. These seldom come up in interviews, but most are pretty obvious within a month or so of starting.

Given the previous points, one excellent way to evaluate developers is to actually hire them on a contract basis for a small chunk of work. Not only does the developer get exposure to the team, internal processes, and the problem space, but they do so for a longer period of time than a standard interview. If done correctly, both parties will know whether they can work well together (or not) by the end of the engagement. While this process doesn't scale, if prior screening mechanisms work well, it can often produce better results than an interview. I've worked at several companies that have done this and they have been some of the best ones. Not only do you get a real feel for how the company works, but you also get enough exposure to potential coworkers that you can truly tell whether you'll get along with them or not.

Summary

It is difficult for many companies to hire well. Whether it is because of the general difficulty of evaluating a candidate's suitability, various situational problems that complicate the process, or simply because evaluating a candidate's personality is simply a hard problem, most companies find the process difficult. While these all sound like problems for the employer, they also represent an opportunity for the interviewee. If you are conscious of the things that make interviewing difficult for the interviewer, you will have many opportunities to stand out from the rest of the candidates. Most people go into interviews thinking only about themselves, their skills, and how they are going to get the job. If you go in instead thinking about the problems the interviewer (and the company) is facing, you will make better decisions during the interview. When you are deeply aware of the problems faced by others, you can often find ways to work together to create a better situation for everyone involved, and this extends to the process of interviewing.

The First Steps for Getting Ready

In this chapter and the next few, we'll talk about what to do to prepare for the interview. This chapter will center on the things you need to do to make the process as painless as possible. In the next chapter, we'll discuss in depth how to practice well before interviewing. The goal here is to practice until the real thing feels easier than the practice.

Search for Common Questions and Problems

Here's a secret that the folks interviewing you won't tell you—interviewers are busy people. The fact that we're hiring is evidence that there is more stuff to do than we can get done. We don't have time to put a lot of thought into creating new interview questions and whiteboard problems. If we feel that the rest of our interviewing practices are good enough, we can just do an Internet search to find some suitable questions and problems. So, it's in your best interest to do an Internet search for questions as well.

© William Gant 2019
W. Gant, *Surviving the Whiteboard Interview*,
https://doi.org/10.1007/978-1-4842-5007-5_3

Start off by searching for "Interview Questions for [framework/language/platform/database]" where the items in brackets are the framework(s), language(s), platform(s), and database(s) that you will likely be working with for your job. For instance, if you are an ASP.NET developer using Angular and SQL Server, you might make separate searches for the following:

- ASP.NET interview questions
- C# interview questions
- Angular interview questions
- SQL Server interview questions
- IIS Interview questions

Make sure you can answer the most common questions you find with these searches, as you probably will be seeing them again. If you have to research the answers, take the time to do so, because this will help you get through the interview long enough to get to the parts where you can truly differentiate yourself. It will also help you a little bit later if you actually practice with a whiteboard. If you see any short coding problems here, hang on to them for later.

This whole thing may feel like cheating. However, the idea here isn't to memorize the answers; the idea is to internalize them. Not only can these answers help you in the interview, but there are pretty good odds that you'll use the information in your job as well. You have tools that can help you—use them.

FizzBuzz

There is a particular type of whiteboard problem that is used a lot and is perfect for early practice with the whiteboard. That problem is known as FizzBuzz. First proposed by Joel Spolsky, FizzBuzz is a very basic coding exercise that shows you at least have the beginnings of thinking like a programmer. By itself, it's not enough to prove that you can code in a real environment, but it is still used to screen job applicants.

Besides being simple enough to start with, FizzBuzz offers a number of other advantages. For one, there are versions for most programming languages, and if there aren't, you can probably still figure it out without much trouble. Second, FizzBuzz problems in another programming language are still understandable, even if your friends don't program in the same language as you do. This makes it easier to find a practice partner for the whiteboard. Finally, FizzBuzz does show familiarity with things like loops, the console, arithmetic operators, and strings.

Again, do an Internet search and get the basic outline of FizzBuzz in your language of choice and set it aside. We'll be using it in Chapter 4 as a code kata.

Code Katas

In addition to practicing whiteboard problems on occasion with a friend, it will help you a lot to practice code katas daily. The main goal with continual practice of code katas is to get you into a headspace that is effective for getting through technical interviews and keep you there. You should also revisit the concept once you have a job, as they can help you a lot when you are trying to learn new skills as well.

Katas in the martial arts are prearranged sequences of movements to help you practice. They help improve a lot of things like footwork, balance, body structure while doing a technique, and flexibility. While not directly helping with getting through actual combat, they are quite good for developing ancillary skills that are still really useful. The same practice can be applied in coding.

Again, to get some reasonable code katas to begin, go to Google and search for "[framework/language/platform] code katas" for each framework, language, or platform that you use. Pick one that looks a little challenging to you (but don't overdo it) and practice it every day. The idea here is to use the same example over and over, working it all the way through and trying to improve your approach. You might think that most of the value of a kata is gone after the first time you complete the problem, but nothing could be further from the truth. After you get done solving the problem, try to solve it faster, more eloquently, with unit tests, using less code, and so on. You can learn a lot from a single problem practiced in this fashion, and the lessons thus learned will improve your skill. Keep practicing until you are sure you have wrung everything out of the kata that you can get, then pick another and do it again. Periodically revisit the katas you've worked through before and see if you've picked up anything new. You might be surprised. We'll spend some time in the next chapter explaining how to really improve your skillset with code katas, but for now just pick one and try out the concept.

This all sounds corny, doesn't it? However, it's a trick I've used on a regular basis over my entire career, both to sharpen my skills and to bring new techniques into my practice. Even in my current role as a software architect, I still periodically go through a period of practicing code katas to tighten up my skills. If you do this as an aspiring developer before you get your first job, you'll rapidly stand out from the crowd. Whiteboard problems are often very similar to code katas, and consistent practice with code katas will make you better at getting through whiteboard problems.

My friend BJ, who is my partner on "The Complete Developer Podcast," among other ventures, was one of the first people I worked with on surviving whiteboard interviews. We drilled whiteboard problems. We drilled code katas. I was tough on him on both—in fact I was far tougher on him than an interviewer would be. When it came time to interview and he was given a whiteboard problem to solve, he laughed. The interviewer picked a problem that he had been working on as a code kata for a couple of weeks. He solved it in a single line of code, as the interviewer stood by, mouth agape. Needless to say, he got the job. Even better, his skills stood up to the test of the job after he got it. We will discuss code katas and FizzBuzz in Chapter 4.

Find a Study Partner

Another thing you need to do is practice for whiteboard tests. It's going to require working with someone else, but it is worth the trouble. Practice actual whiteboarding problems with a friend, as often as possible, until you actually manage to land a job.

Each time you practice, each of you will present the other with a problem to solve on the whiteboard. When it is your turn, you will solve the problem as if it was a real interview. You'll get some advice on how to get through a real interview in a later chapter. Practice on the whiteboard with a friend who is acting like an interviewer. This is not a joke, so be serious. Your goal is to practice in such a way that when you encounter the real thing, you are ready.

When it is your turn to be the interviewer, your job is to be a challenging interviewer. You'll have to vary this depending on your partner's tolerance, but your job is to press them to the point of mild discomfort. Question what they are doing, and why. Make noise, joke, talk on your phone, and do all the stuff that you would imagine a terrible interviewer doing. Give your partner the opportunity to practice under pressure, especially irritation and emotional pressure.

When I worked through this with BJ, I was a jerk toward him, on purpose (I had a blast though). I can remember one day when he made a mistake, when I threw an eraser at the back of his head, telling him to pick it up and erase the memory leak he had just written. While I don't necessarily recommend taking it that far (we'd known each other for 15 years at that point), I will tell you that getting through such an experience in practice makes it far easier to get through it in a real interview, when the interviewer is probably far nicer. Practice hard, so that the real thing is easy.

The idea behind doing this is that exposure to a harder version of a whiteboard interview with nothing at stake will make it easier to emotionally and intellectually handle the real thing when your ability to pay the rent is on the line. You don't want your first experience with a whiteboard interview to be during a real interview for a job that you actually want. Take the pressure off in advance by practice.

Write Code Every Day

I almost hesitate to mention this, but you really do need to try and write code several times a week, if not daily, while you are interviewing. It's not just that practice is important, but that your skills will atrophy faster than you think if you aren't using them regularly. If you're lucky, your job search for your first development job will only take a couple of months, but it might end up being six months or longer.

I've worked with a lot of people who have recently graduated from coding bootcamps, undergraduate programs, or have simply learned to code at an acceptable level by themselves. The most common reason for failure of recent graduates to get a software development job isn't lack of knowledge, it's atrophy of knowledge while they're waiting. Like anyone else, interviewers usually see what they expect to see, unless you give them a reason to look more closely. Regular software development, in addition to improving your skill, also hones your reactions and thought processes so that you "look" more like a developer. Atrophied skills make you stand out, and not in a good way.

Once you get a full-time developer job, you're probably still going to have to learn on your own time. While employers provide varying levels of training, it's rare that you're going to be able to get enough training during work hours to keep your skills fresh. Since it will help you and you're going to be expected to do so anyway, you should continue working to build your skills.

There are a number of different ways to build your skills, including going through more coursework. While coursework helps to a degree, the best thing an aspiring developer can do is to build a small app that does something useful. You can do almost anything (I think I've built a little recipe management application in 10 or 12 different platforms over the years), but try to pick something that has some of the same kinds of problems that you expect to encounter in a job. In general, I would recommend choosing something where you have to deal with databases, authentication, and authorization at a minimum. This will give you a lot of moving parts to worry about as well as force you to deal with insufficient documentation, trying to figure out the best way to do various things, and open-ended design. If you are already working on code on a regular basis at home, you'll be able to keep your skills fresh enough for interviewing to go well.

Summary

In this chapter, we discussed general things that you should be doing to prepare yourself for interviewing. These tactics are not company-specific. Rather, they are designed to sharpen your skills so that you can handle any interview. We discussed how to use a search engine to get some basic information you need for practice. Next, we briefly went over what FizzBuzz is, how to find it, and what it can do for you. We followed that up with a discussion of practice whiteboarding and the importance of continuing to write code for practice. In the next chapter, we'll be digging in more on how to do code katas and some possible approaches to really ramp your skills up to the next level using them.

Code Katas

In the previous chapter, we discussed using code katas to help prepare yourself for an interview. However, there is a little bit more to it than just simple preparation. In this chapter, we'll discuss some ideas for fully unlocking the potential of code katas. If you do code katas well, you can not only prepare for an interview more effectively but you can also improve general coding skills. In this chapter, we'll explain a simple code kata, using the FizzBuzz coding exercise.

I personally use code katas regularly to improve skills in various areas. Because they are usually simple practice problems, you can easily fit them into your schedule without much effort. Because I use C# professionally, I've done the following with code katas. You may have to adjust these ideas for your own software stack, but they might give you some ideas of your own. Use code katas to make the following improvements in your software development:

- Implement test-driven development practices or TDD.

- Learn to use the Task API and the async primitives.

- Do the whole thing without using a mouse.

- Write the code in fewer lines of code, or even with fewer characters.

- Solve the coding problem in less time.

- Use a different language to solve the coding problem (I've done C++, JavaScript, Ruby, and Python).

- Learn to use a different IDE for the same language.

- Learn to use new IDE extensions and plug-ins for part of the process.

© William Gant 2019
W. Gant, *Surviving the Whiteboard Interview*,
https://doi.org/10.1007/978-1-4842-5007-5_4

- Practice debugging shortcuts.
- Use a different Object Relational Mapper (ORM), or database server.
- Use a different front-end library to accomplish the same task.

You can really squeeze a lot of value out of a code kata besides just learning how to solve a particular problem. In particular, code katas are a nice way to familiarize yourself with your development environment and the various tools that you are using, without the burden of a large and ugly project in the mix. Additionally, they can help you test out a new library quickly, instead of trying it in your main application, where the work is liable to take far longer.

FizzBuzz

In a moment, we're going to go through the most basic whiteboard problem out there. That problem is called "FizzBuzz" and is the problem I mentioned in Chapter 3. Essentially, solving the problem shows that you have a basic understanding of how to think like a programmer (and not much else). In the next section, I'll walk you through the process of completing this exercise in a number of different common programming languages. Don't worry if your language of choice isn't represented—the point is to understand the process of working through this problem, not to get hung up on the language.

The General Idea

Essentially, the FizzBuzz problem is to write a function that does the following:

- Loop through the numbers between 1 and 100
- If the number is divisible by 3, print "Fizz"
- If the number is divisible by 5, print "Buzz"
- If the number is divisible by 3 and 5, then print "FizzBuzz"
- Otherwise just print the number

This process shows that you understand how looping constructs and flow control statements work. It also shows that you understand things like the modulo and equality operators, outputting text to the console, and string manipulation. Pay special attention to how this code was formatted. I tried to make it idiomatic for the language. This serves two purposes. First, it makes it easier on the interviewer to read your code, and second, it doesn't immediately irritate long-time users of the language in question. You should try to do the same when you are working through whiteboard problems, for the same reasons.

C#

In C#, a developer might be expected to solve this problem as follows:

C# Code

```csharp
public void FizzBuzz()
{
  for(var i = 1; i <= 100; i++)
  {
      if(i % 15 == 0){
        Console.WriteLine("FizzBuzz");
      }
      else if(i % 5 == 0){
        Console.WriteLine("Buzz");
      }
      else if(i % 3 == 0){
        Console.WriteLine("Fizz");
      }
      else{
        Console.WriteLine(i.ToString());
      }
  }
}
```

C++

The way a C++ developer might handle this is fairly similar to C#, with the obvious differences in the way console output is done and variables are declared. Note that this assumes that you have imported the standard namespace (std).

C++ Code

```cpp
#include <iostream>
using namespace std;

void FizzBuzz()
{
  for(int i=1;i<100;i++)
  {
    if(i%15==0)
    {
      cout <<"FizzBuzz" << endl;
    }
```

```
    else if(i%5==0) {
      cout <<"Buzz" << endl;
    }
    else if(i%3==0) {
      cout <<"Fizz" << endl;
    }
    else
    {
      cout << i << endl;
    }
  }
}

public int main()
{
    FizzBuzz();
}
```

PHP

In PHP, you might have the following function body. Note that instead of outputting to a console, PHP folks are more likely to output to a web page. They do this because it's something PHP is particularly good at. It's right there in the name (**PHP: H**yper**T**ext **P**reprocessor—they used a recursive acronym).

PHP Code

```php
for($i = 1; $i <= 100; $i++)
{
  if($i % 15 == 0){
    echo "FizzBuzz<br/>";
  }
  else if($i % 5 == 0){
    echo "Buzz<br/>";
  }
  else if($i % 3 == 0){
    echo "Fizz<br/>";
  }
  else {
    echo $i."<br/>";
  }
}
```

Ruby

Ruby has a lot of things to help you make your code shorter. If you can fluently whip these out during an interview, it'll go a long way to showing your skills. I don't use Ruby in production, but this sort of stuff is why I'll always have a soft spot in my heart for this language.

Ruby Code

```ruby
(1..100).each do |n|
  a = String.new
  a << "Fizz" if n%3 == 0
  a << "Buzz" if n%5 == 0
  a << n.to_s if a.empty?
  puts a
end
```

JavaScript

In JavaScript, one might solve the problem thus. Note the triple equal sign. Also note the use of the "let" keyword. This is for ECMAScript 6. If they are using an older version, this would be "var" instead, so be sure to ask. If you don't know why the triple equal sign is important or how "let" differs from "var," be sure and take some time to look that up and get your head around it.

JavaScript Code

```javascript
function FizzBuzz(){
  for(let i = 1; i <= 100; i++){
    if(i % 15 === 0){
      console.log("FizzBuzz");
    }
    else if(i % 5 === 0){
      console.log("Buzz");
    }
    else if(i % 3 === 0){
      console.log("Fizz");
    }
    else{
      console.log(i.toString());
    }
  }
}
```

Python

Python is also a beautiful language that I don't get to use in production. Like Ruby, there is an eloquence to the code that some of these other languages miss. However, in this case, you might not want to get too slick with it.

Python Code

```python
def fizzbuzz():
    for n in range(0,100):
        output = ""
        if(n % 3 == 0):
            output += 'Fizz'
        if(n % 5 == 0):
            output += 'Buzz'
        if(output == ""):
            output += str(n)
        print(output)
fizzbuzz()
```

Summary

In this chapter, we discussed some practical ways to use code katas to improve your skills. We first discussed some particular areas that can be improved by judicious use of code katas. Next, we worked through a very simple code kata that might also show up when you are given a whiteboard problem during an interview. While this was a very simple problem, it's exactly the kind of problem that can be a perfect practice problem for improving your development skills.

Your Resume

In this chapter, we're going to briefly discuss how to think about your resume. There's a lot of debate on the "proper structure of a resume" on the Internet. There are a wide variety of opinions and they shift a lot over time, more often resembling fashion advice than the sort of "engineering and math" advice that is more typical of a development career.

This was probably the hardest chapter in this book to write. It wasn't because of the availability of opinions, which are too numerous to count, but rather because every opinion made significant assumptions about what constituted a "good" resume. Most of the time, the descriptions of what constituted "good" were as highly subjective as you might expect from an average person's description of a "good dinner," "good vacation," or "reputable news source." In short, while there are some reasonable guidelines on these things, the sheer volume of purely subjective junk sitting on top of them makes a lot of the advice useless. I've read articles about how you need a one-page resume, how the one-page resume is dated, why you need a CV instead of a resume, and even how your work products are effectively your resume (I tend to agree with this last one). They'll still be debating this in a hundred years, but almost all of us will be looking for jobs in the meantime.

Instead, what you really need is a way to subjectively create the best resume you can for yourself, and have it be "as good as possible" for you, based upon your own goals. To be honest, not many people care too much about their resume, so long as it provides the results they want and leaves some options open. While there are a few competitions for resume writers (such as the National Resume Writers' Association ROAR contest), the actual events are unlikely to ever be televised or to win anyone any fame. As a neighbor of mine put it when I was growing up, "you don't want to be the best one at

© William Gant 2019
W. Gant, *Surviving the Whiteboard Interview*,
https://doi.org/10.1007/978-1-4842-5007-5_5

shoveling out the horse stalls—you want to get the job done so you can go ride." Odds are good that your feelings about spending excessive amounts of time in your favorite word processing software are the digital equivalent of that old farmer's sentiment (and, more importantly, correct).

So, what do you do? To start with, you don't start by thinking about what your resume "should look like." Instead, shift your thinking to what it "should accomplish" for you and how it should set the tone for your upcoming interview. It's unlikely that someone will hire you for any job that you actually want based solely on the strength of your resume. Getting your resume right is simply the first step in a big long staircase that leads you to have a successful job where you are content. The goal of the resume is to get the hiring manager to take the next step. Beyond that, your resume might have a little use for discussion during the actual interview, but it is not going to get you hired on its own.

In this chapter, we'll be discussing some general rules around the construction of your resume. While resumes are important and required, they really don't do as much for you as you might be led to believe. Think of your resume as being more of a marketing document than an overly detailed account of events, and you'll probably be better off. This doesn't constitute an excuse to falsify anything—it's just that an overabundance of detail is probably less useful than something smaller that actually moves the process forward.

Why a Good Resume Is Important

You've probably heard it said that first impressions are really important. This is something most people agree with, as most of us have had the experience of having to overcome an initial bad impression. However, most job seekers don't consider that the first time a hiring manager learns anything about you, that they probably learn it from your resume. Your resume is the business equivalent of an introduction at a dinner party and, as such, needs to be carefully crafted so that it gives the best impression possible.

Many people have the view that their resume should simply be a list of their previous jobs and accomplishments. While a resume should definitely contain those things, a resume that contains only those things is not particularly useful. A resume has two major goals:

1. Get the attention of the hiring manager such that they actually want to start a conversation with you.

2. Help set the tone of the interview in a way that favors your strengths at fixing the problems the business has.

This sounds simple at first, but good execution of these goals varies a lot based on what experience you have, the type of job you are looking for, and the desires of the hiring manager. No matter how good you are and how much experience you have, if your resume is only a listing of your previous jobs, you are missing out on some opportunities to impress the interviewer.

Instead of just explaining what you did in a previous job, there is a better way to express your experience that actually meets the goals we listed earlier. The remainder of this chapter will be about how to do exactly that.

Essentially, every step of the interview process, including this one, is about trying to move to the next step of the process, while laying the groundwork for that piece to be successful. Your resume is part of the interview, whether you like it or not. Let's make it count.

Tailoring Your Resume

For a variety of reasons, you will need to adjust your resume based on where you are applying. Different industries have wildly different viewpoints as to what constitutes a "good" resume. In addition, your interviewers are going to be looking for different skills and experience. Finally, what is considered "best practice" in regard to a resume tends to change over time. It's not quite as fad driven as your average front-end JavaScript framework, but it is close enough to be annoying.

As a result, almost any direct advice I can give you in regard to formatting your resume is quite possibly pointless and very likely to be out of date by the time you have this book in your hands. In addition, it may not be appropriate for the industry you are targeting. An interviewer at a large accounting, legal, or banking firm will not have the same expectations as a small startup, marketing, or media company.

Instead, what I suggest that you do is prepare a version of your resume that has everything on it, and then make a copy of it and edit it for your intended audience. These edits should take the following into account:

- General tone and language used. You describe things in a much more formal fashion for an older, stodgy company, whereas you might be a little looser with your language for a startup.

- You should change color, fonts, layouts, and the like to match the expectations of the target market. Again, you'll want to go with more subdued colors and fonts when being interviewed by more "old school" organizations, but you will make a better impression with many startups by being more relaxed.

- You should consider the use of graphics when appropriate.

- You may wish to remove some of your experience from the list, particularly if it is irrelevant or gives an impression that you don't want. For instance, a lot of the contract work I've done has been with smaller companies and startups—I wouldn't want that to be too prominent if I was trying to get a job with a larger company in many cases. They might well think that I wouldn't be happy there or would have sloppy habits from being in less formal environments.

Above all, you should remember that a resume should be more like a personal letter, rather than junk mail. If you really want a high-quality job, your goal should be the development of quality interest in your skills, rather than simply blasting your resume out to everyone who might be interested. I've sent my own resume out before without customization, and I can tell you that the results of doing so are lackluster unless you get extremely lucky.

Always keep a copy of your full resume handy. You'll be tweaking and expanding this over the years, and you are far better off not having to recreate everything on your resume when it comes time to look for another job. In addition, try to have a recurring calendar reminder to update your resume every 3 to 6 months. Not only will this make sure that you don't forget any recent accomplishments, but it will also make a job loss much easier to recover from.

Over time, you'll also find that your career goals tend to change. When I first started software development, I wanted to develop software for manufacturing, simply because I thought it was neat. As I've progressed through my career, I've come to realize that manufacturing doesn't perfectly match what I want to do, as I also want to do a lot more work that is visible to the public and directly subject to market forces. As a result, I've reworked my resume a number of times to highlight things that will be useful in that direction while drawing less attention to things that aren't.

There is one final caveat on your resume that might be of use. When you are posting your resume on a platform like LinkedIn, it can be difficult to appropriately tailor for each individual organization where you are interviewing. In addition, if you are like me and doing several things at once (developing, managing, doing a podcast, running meetup groups, writing books, etc.), it's completely impossible to create a well-crafted resume that pleases everyone. In this case, don't do that. Instead, put the full resume up there. In this case, you don't want to leave anything out, because you can be almost certain that people are finding your resume because they searched for certain words.

How to Word Things

Instead of just a listing of all the jobs you've had over the course of your career, you need to rephrase your accomplishments in such a way that they highlight the value that you bring to a team. For instance, compare the following two lines from a resume:

- Helped convert the payment system from a monolith written in classic ASP to a microservice architecture using Ruby on Rails.

- Helped convert a payment system that was handling one million dollars in transactions weekly from classic ASP to a microservice architecture using Ruby on Rails that allowed the company to scale to four million dollars of sales in a week.

While both versions will probably be interpreted in the same way by a developer, the way that they are interpreted by a hiring manager will be wildly different. While development skills are required to be successful in a development job, they are far from the only skills required. Additionally, developers have a tendency to fixate on programming languages and technology stacks as measurements of their value. However, the technology is only a means to an end—you can build a perfectly acceptable application using tools that aren't popular.

The real crux of the matter is that being able to code effectively is nothing more than a baseline, even though you will encounter lots of developers who believe that it is the only thing that matters. Whether the organization is a small business, a large corporation, or a government, they aren't really hiring for software development skills; they are hiring so that they can fix problems. While many of the problems you'll be fixing will require software development skills, that isn't true of all of them. In fact, the further along you are in your career, the more likely it is that you'll spend a substantial portion of your time dealing with things like specifications, personnel issues, and dealing with stakeholders.

Further, being able to quantify the value that you provide makes it much easier to negotiate for a higher salary or to argue that you should be in a more senior position in the company. If someone is "just" a developer, their skills don't support the same kind of salary that can be expected of someone who is a developer and knows why they are writing code. If you show that you understand how your work contributes to revenue, it changes the first impression that you make with interviewers.

Essentially, the way you word things in your resume can greatly impact the first impression that you make with an interviewer. That first impression can not only get you in the door but it can also change the entire tone of the interview. Instead of being "Generic Ruby Programmer #47," you are "Cynthia, the developer that reworked that payment system. I wonder if she can help us scale our inventory control system." The latter is a far better place to find yourself, because it both helps you stand out and gets the interviewer thinking about exactly how you can help them before the interview.

In addition to making sure that you present your experience in the best way possible, you also need to make sure that you don't present information in a way that harms you. Be especially careful how you describe problems that were occurring at previous employers, as it is really easy to accidentally give more information than you intended (especially if you are interviewing with competitors). If you do this, companies will be concerned about hiring you, especially if they are worried about competitors obtaining sensitive information about the company.

You should also try to express things with as few words as possible while remaining accurate. Hiring managers and human resources personnel who read resumes tend to skim resumes, especially if they receive a lot of them for a particularly desirable position. You are better off making it easy for them to scan and find what they are looking for.

You should also make sure that critical keywords are in your resume. In addition to skimming your resume, hiring managers may also simply look for certain keywords, tossing your resume in the trash if they don't find them. Sometimes they even use software to do this.

Remember that the words on your resume don't only have to tell a story about your career, but that they also have to convince someone that it is worthwhile to follow up with you. The way you frame your achievements on a resume can make the difference between getting an interview and being passed up.

Summary

In this chapter, we discussed some general rules around how to structure your resume so that you can get the attention of the hiring manager long enough to be called in for an interview. In addition, we discussed how careful setup of your resume can change the tone of the interview in your favor before you walk in the door. Finally, we discussed some general things you might want to change when customizing your resume for different audiences. In the next chapter, we'll discuss ways that you can cultivate references and build your personal network so that you continue to make a good impression before walking in the door.

References and Your Network

Long before you start interviewing, you should be building your network and finding good references. Whether it is for your first job or your twentieth job, employers are likely to want to hear from your references. Good references can do anything from simply proving that you worked somewhere to providing references for your good character, work ethic, and skill level. Even for your first development job, it can often be very worthwhile to have a number of good references to vouch for you. In this chapter, we'll discuss the process of networking and getting good references. A good network is both helpful for finding job opportunities and for finding people who can serve as references.

Why Networking Is Crucial

You may have gotten the idea that programming is a solitary profession, populated by "hacker" loners who do everything from a basement somewhere without interacting with anyone else. While that's a popular TV trope, it hardly resembles the life of most successful programmers. Sure, you can get a certain level of skill and maybe even a job or two, just by yourself, but being a true loner really makes things more difficult.

The fact of life is that nobody pays you anything for anything if they can't find you. While you may have an elite level of skills, if employers can't find you, they aren't going to pay you for them. In addition, most programming projects

© William Gant 2019
W. Gant, *Surviving the Whiteboard Interview*,
https://doi.org/10.1007/978-1-4842-5007-5_6

that can be completed by a single person have been completed by a single person already. You'll be hard-pressed to come up with a project that you can do all by yourself. In reality, you not only have to work with other people but you have a distinct and sizable advantage if more people know who you are and have a good impression of you; it makes it a lot easier to find both the first job and subsequent jobs. Furthermore, if people like you enough, they may seek you out with potential job offers when you aren't even looking. I've gotten some of the best jobs I've ever had because someone thought of me when their company was hiring.

The value of a good professional network simply can't be overstated. Besides the job prospects, it offers opportunities for mentoring, for learning about new technology from people already using it, and can allow you to find out which companies you don't want to work for in your local area. In all honesty, a good professional network can often exceed the value of high skill in many cases.

Local User Groups

Your first and best choice for starting to build up your professional network is to find one or two local user groups that deal with your chosen language and platform and to attend their meetings with some regularity. The best place to find potential meetups currently is Meetup.com. You may have to work your way through a few groups to get to one that actually meets your needs. When picking a group, bear the following criteria in mind:

- Does this group meet consistently, at a time and location that allows me to do the same?

- Are the members friendly and accepting of new people?

- Do the members have a good range of experience? Are there people there more experienced than you? Are there people there less experienced than you?

- Are the meetings well organized?

- Are the topics of the meetings actually useful?

You probably should attend two or three meetings for each group in order to be able to evaluate these effectively. In particular, the first item and the last two items are difficult to determine with just a single meeting. The third item may be a little counterintuitive as well. In general, you don't want to be the least experienced person at a meetup, because it may make you feel awkward when you ask a question. In addition, if you are a little too new for a particular meetup, you are often better off finding another group until your skills improve.

Also, be sure and avoid any groups that seem to be unfriendly to outsiders, disorganized, flaky with their schedule, and the like. Having a consistent group to work with will greatly increase your odds of success, while an inconsistent, unfriendly, or disorganized group is an excellent way to ruin your enthusiasm and waste your time.

When attending a group, you need to make an effort to network with the people there. While there are entire books available on suitable networking techniques, networking for software developers is much easier than the sort of networking required for things like sales. Mainly, you need to introduce yourself to the other people in the meetup, paying particular attention to people that you don't already know. In general, this works better if you try to find out what people are doing, rather than trying to explain what you are doing. In the early days, you'll learn more on average by finding out what someone does and asking questions than you will by describing what you are doing and then asking questions.

When you leave, take time to write down appropriate details about the people you met, along with connecting to them on professional social media such as LinkedIn. When you are looking for work at some point in the future, you'd be surprised how often some random person that you met at a meetup group will be the one that tells you about a possible job. The first job is really the hardest to get in this industry, and having a good, established network is critical if you currently don't have a long history of successful projects in a professional environment. Even for experienced developers, regular cultivation of a personal network is extremely useful for finding better jobs.

Open Source Projects

Another option if you are unable to attend face-to-face meetups (or if you wish to supplement that effort) is to participate in open source projects. It's a fairly trivial effort with most search engines to find a list of open source projects in whatever language or platform you want to use. Open source is a practice that is here to stay and is a great way to get some exposure to people that might want to hire you, but you do need to pick a tool that is likely to be used in the sort of environment in which you hope to work.

However, there are few open source projects (or software projects in general) that are as straightforward and simple as the sort of projects you do while learning a new language. Actual production-level open source projects tend to use more advanced concepts like plug-in models, dependency injection, automated testing, and build scripts. As a result, it can be difficult to find something to contribute that is both useful and that you are capable of completing without annoying the project maintainers.

Instead of trying to make major changes to the software, fix bugs or even write documentation, probably the easiest thing you can do is to find the bug list for the project and investigate a few of them. You can then point out your findings on the discussion thread for the bug in question. This is a fairly low-impact way to build some rapport with the project maintainers. After doing this for a while and building up some understanding of the codebase, you might be able to contribute to the project by fixing small cosmetic bugs or by writing documentation.

In addition, if you have a working blog (as described a bit later), you can also write tutorials on using the open source tool there. While not part of the official project documentation, well-written personal blog posts can still help establish you as a knowledgeable resource. Even though you may be eager to get started, don't rush the process of jumping into open source at an early stage. You'll make a far better impression with the project maintainers by a measured, long-term approach than you will by trying to jump in and fix everything from the start. While this seems counterintuitive, you must remember that the maintainers for an open source project are responsible for the stability of the project over the long term. As a result, they are frequently reticent to allow new developers to simply jump in and make changes, especially when those developers don't have a consistent track record.

It takes a while, but a good track record with open source can really help set the stage for a successful interview, especially if the interviewing company uses that product (or something similar to it). If you show yourself to be experienced with the software, the interviewer will expect you to be more knowledgeable and will probably treat you differently than they would a brand new developer without experience. In particular, this tends to make the interviewer start trying to determine what you know, rather than what you do not. It doesn't seem like much of a difference, but it can really change the character of the interview.

Open source contributions can also help you get the attention of potential employers. This can result in the employer reaching out to you about an open position, rather than you approaching them. This also significantly helps the tone of the interview since they reached out to you.

Former Coworkers and Friends

One thing that many aspiring developers forget when trying to improve their personal network is that they already have one. In all honesty, unless you've been living under a rock, you likely already have a personal network of hundreds of people. I grew up out in the sticks, at the end of the power line (literally), yet just the contacts from my childhood comprise hundreds of people, dozens of whom have careers or contacts in computer-related disciplines. And this is from a guy who grew up in small town Tennessee. I later went to college in the "big city" (Nashville) and met thousands more people.

The point is you already have a network and you are absolutely insane if you don't use it to your advantage. Even if none of your former coworkers, friends, and contacts from school are professional software developers, it's highly unlikely that the same is also true of all their friends and contacts. In addition, if you are thinking about becoming a contractor or consultant, your nontechnical contacts can also help you out to a surprising degree, either as potential clients themselves or because of their own networks.

At this point, I'd also like to mention that most of us tend to think of old friends and coworkers as they are when we were around them, rather than remembering that they change as much as we do. As a result, it's quite possible that even if you don't think you know someone in this industry (or someone who can otherwise help you), you actually do.

Now, you might be tempted at this point to start sending emails to people. Don't do that. If you know even a moderate number of people, sending emails to all of them will quickly get out of hand. Not only will it take a bunch of time to send emails to everyone, but the responses will also take up a lot of time. In addition, you may not want everyone to know that you are in the process of trying to get a software development job. I don't know about you, but just because I know someone doesn't necessarily mean that I want to trust them when I'm trying to start a new career. It also doesn't mean that they are actually able to help me in a useful fashion (for instance, I have multiple friends in Australia who would love to help me find a job, but I don't plan to move). Instead of contacting your entire list, you should segment it and tailor your messages based on what you can do for them and do it slowly.

It may seem a bit annoying, but the fastest way to get someone's attention is to be useful to them. When your approach to other people is based on trying to get them to help you, it's very easy to turn people against you or simply be ignored, whereas if you have the potential to be useful, people are more likely to help you out. "Useful," in this case, doesn't necessarily mean that you do work for them (although if they have simple computer problems you can help with, that will still help you). Instead, get into the habit of making contact with a few people a week, and take time to actually find out what's going on with them. Keep notes (I use a spreadsheet when doing this), and forward things on to them that you think might be useful to them. This can include jobs if they are looking for one or potential business contacts for their employer. Ideally, what you want is to have a list of people to whom you are a valuable contact. This can be especially helpful if you connect members of your network who need things with other members of your network who provide those things. That's the first step to getting on people's radar and staying there.

There is one other thing you can do for people in your network that is valuable and doesn't necessarily require you to find them business contacts or jobs, and that is to serve as a reference for them. Generally speaking, if you've had at least some work experience, you will have had a positive working relationship with at least a few people. You should reach out to those people and let them know that you are happy to serve as a reference if they ever need one. In addition, you can write a recommendation for them on LinkedIn or other professional web sites. Besides helping the competent people you know find work, this also gets you into a good habit, that being the habit of serving as a resource for other people. While it may seem like a small thing now, this is a habit that will pay off well further along in your career.

Now, I know that at least some readers will read the preceding items and think that it sounds a little like I am suggesting that you should help people so that they will reciprocate. It sounds that way, but it isn't. You should be doing this without the thought of reward, but that doesn't mean it's out of a lack of self-interest. In essence, the best way to think about it is that you are a part of an ecosystem and these practices help improve the ecosystem. While your career is your own responsibility, it will be infinitely improved by having a solid network of people around you. The experience of the other people in your network will similarly improve by your efforts. Instead of looking at yourself as being separate from others, it is far better to think of yourself as a representative of a larger set of people, all of whom are going in the same direction. This mindset will take you further than simply going it alone.

Internet Punditry

Given that you are looking at yourself as merely a representative of a larger group of people, you also need to start spending some effort to expand your reach. The easiest way to do this is to participate in communities on the Internet that match your intended goal. Besides the obvious benefit of establishing yourself along with a community of people around you, establishing yourself in a community will also help accelerate your learning. There's nothing worse than being stuck on a simple problem for hours (or even days) with no help in sight. Participation in a community can help protect you from this.

However, before you start participating in a community, you might want to consider a blog to chronicle your journey. While the process of creating a blog is outside the scope of this book, there are a number of web sites that will let you set up a blog for free. It doesn't have to have a lot of features, but be sure and select a blog host that makes it easy to post source code and images. Don't worry too much about this step—if you don't pick the right site to host your blog, you can change it later. The important thing to do is get started. You don't even have to make the first post about code; in fact, that's probably not the most useful thing you can do. Instead, gradually introduce yourself and your struggles in a blog. Write a little every day about the things you learn as you go. Not only will it be something that you'll want to look back on, but it will also help readers invest themselves in your story.

At this point, you may be concerned that a blog will draw attention to yourself. If you are like I was, you'll resist shining the spotlight on yourself. Whether it's because the open Internet has some really messed up people on it (it does), or because people like to troll random strangers on it (they do), or because some people won't like your blog (some won't), you should do it anyway. Putting yourself out there on a consistent basis will get the attention of a surprising number of people, some of whom are hiring. If you don't let them push you around, the trolls won't matter. The haters will never go away, but you'll eventually realize that they don't matter. But you have to overcome the fear of experiencing them. There is literally nothing you can do that will be more effective than prominently putting yourself out there. While this is entirely optional, I really can't recommend it enough.

Assuming you've taken the really good advice in the preceding paragraph and are starting and maintaining a blog, now is the time to start looking for a community or two in which to participate. While it varies depending on which language or platform you intend to use, a simple Internet search will usually help you find a few forums in your target language. However, because a lot of web sites play games with search engine results, you may need to look through a dozen or more sites to pick out a few good ones. You'll primarily want to look for sites that have a fair amount of recent activity and where the community is generally positive. In particular, you're going to want to avoid sites that have a lot of trolling and negative behavior.

There are a few other options that you might want to consider as well, particularly if writing isn't your thing. You should also consider podcasting or video blogging if you tend to do better with speaking on camera or into a microphone (I do the latter myself). While these are more complicated to host and can be more expensive and difficult at first, you will find that if you do it right, you can build a fairly decent audience as long as your content is good.

Finally, no matter what you do, you need to be consistent. If you are just participating in forums, you need to do so on a fairly regular basis. If you are blogging, you should try for at least one good post a week. The same is true of podcasts and video blogging. The biggest thing that grows your audience in any of these cases is giving them something that they want on a consistent basis. If you do that, you can get people to continue to show up to consume what you are doing, week after week.

All of this may sound like a lot of effort that isn't related to development at all. That's because it isn't. It is, however, a way to get other people invested in your personal journey into software development. It makes you stand out from the crowd of people who are just learning to code and not doing much else. While it sounds a dumb marketing trick, it really works well to humanize your journey into development, rather than you simply being another anonymous face out on the Internet. Further, it provides a body of consistent work that shows that you are serious about this. There is no substitute for getting yourself out there.

Like open source contributions, regular production of blog posts, podcasts, or video will draw attention to you, some of whom are looking to hire developers. Many of these people will already be very familiar with you through your work, and that can work to your advantage when you are interviewing. There is a lot of credibility to be gained by creating material that other people use. In addition, it will give you practice at articulating your ideas, which comes in handy when you are doing development work and need to express your ideas to a team.

Getting References

A good network is helpful for finding out about available jobs, but it can do a lot more for you. One of the most critical things your network should do is provide you with references.

At some point in the hiring process, companies will check your references to determine whether they want to work with you. This often happens even if you are just doing contract work with them. Previous employers often give very limited references due to liability concerns, with many of them simply confirming that you worked with them and when you were there. Obviously, if you can get better references, that makes the rest of the interview process much easier. Good references help you stand out from the crowd, and you should be trying to cultivate as many as you can.

You should begin reaching out to previous coworkers, as well as people that you have helped through user groups and open source projects. However, unless you know the person really well and have remained in contact, it might not be the best time to ask for a reference. Instead, it's often better to ask them if they know anyone that is hiring developers. Not only will this give you good information about open positions, but many people will offer to write you a reference if required. If they don't openly offer to write your recommendation, you should ask if they can introduce you to anyone at the company. Because they offered the introduction, the company is likely to contact them for a reference anyway.

In addition, you should also be writing references for other people. Your career lives within a larger ecosystem that includes the careers of the people in your network. The better off your network is, the better off you are. That doesn't mean you should be writing references looking for a reward, but rather that you should be looking out for your network. You'll find over a long career that many of the same people show up over and over again. Sometimes you write references for them; sometimes they write references for you. I have a friend that hired me at one job, and I recommended him for another job a few years later. Next, he wrote a reference for me for another job. Two contracts later, I wrote one for him and he joined my team. Unless you are truly in a huge tech market, this will probably happen for you at some point as well.

Additionally, it's valuable to write references for others because it makes you more aware of what constitutes a good reference, what you want people to say when writing a reference for you, and how to clearly express these things. Not only does it make glowing references better, but it gives you a good idea of what you should be doing to get better references at your next job.

When writing references for others (or asking for them for yourself), there are a few things that need to be present:

- You need to express how long you've known the person and in what capacity. This makes your reference more credible.

- You need to keep it positive. Nearly everything can be spun in a positive light. For instance, I've worked with good developers who were argumentative. In a reference letter, you either should not mention that, not write the letter at all, or turn it into a positive. While the argumentative types can be a pain to work with sometimes, they also force you to keep things honest, especially if they know how to argue well without making it personal.

- You also want to make sure that the reference letter is targeted toward the kind of job they are currently pursuing. For instance, if you knew the person as a junior developer, but they are now moving into the role of a database administrator, you need to shape the content of your letter in a way that helps them get that job, instead of another junior developer role.

Recommendation letters and other forms of endorsement are not hard to write, but you need to practice doing so in order to do them well. This practice is critical, not only for helping your network to find better jobs but also to learn what kinds of things you should be looking for when others are writing recommendations for you.

Summary

In this chapter, we discussed how and why you should always be working on expanding your personal network. We started by discussing how a good network is critical to a successful career. Next, we explored how to build your networks by participating in user groups and open source projects. Then, we discussed the process of reestablishing contact with former coworkers and friends as part of your preparation for the job search. Participation in Internet-based communities and building a blog were two more ways to expand your reach that we explained. Finally, we discussed how to ask for and give references as part of building the value of your personal network.

How to Clean Up Your Online Persona

In this chapter, we're going to discuss what you need to do to make sure that your Internet presence doesn't keep you from getting an interview. While the Internet can definitely help accelerate the process of getting a job and establishing credibility, it also has a downside. Many of us grew up with at least some presence on the Internet during our "less prudent" years. The younger you were when the Internet became commonplace, the more likely it is that you have said or done something foolish that is forever recorded there.

Some Examples

It's very easy to use the Internet badly and have that come back to bite you later. The following are some pretty simple things that can make you look bad to a potential employer. While some of these don't sound like things that would cause a problem, they can work to your disadvantage in a number of ways.

© William Gant 2019
W. Gant, *Surviving the Whiteboard Interview*,
https://doi.org/10.1007/978-1-4842-5007-5_7

- Forum posts made under your real name and containing inappropriate material. This can be language, pictures, or even heated arguments. While you probably have enough sense not to do this kind of thing in a work environment, the employer has no way of knowing that.

- Forum posts of an innocuous nature in groups that are no longer considered innocuous.

- Microfiches and older nondigital materials being brought online. This process is ongoing and can result in all kinds of fun stuff showing up in searches.

- Public, openly available social media posts of the same nature. While social media accounts are private to varying degrees, if they can be found through a search engine, they will be. If there is anything negative in there, it can be used against you.

- Publicly available criminal records, court proceedings, and interviews with media. If you have a criminal background, it may be harder to get a job, simply because it will show up in a Google search. The same can be true of messy legal battles or interviews with the media where you are quoted by a journalist (if negative).

- Negative posts by other people that have your name in them. Unfortunately, we can't control what our friends (or enemies) do. Sometimes other people create situations that are difficult to handle, publicly available, and preserved forever on the Internet. Worse still, these aren't youthful indiscretions that go away once cleaned up. People can cause you problems at any time.

- Results of data breaches of private information coming online. This can be everything from credit card information, to the contents of private forums, to membership details from sketchy sites.

- People with the same name can also do things that you probably don't want to have on the first page of search results. For instance, someone with the name "William Gant" has been convicted of assault in various places, while another William Gant is a politician in Louisiana. Still another was a guy that got murdered in the TV series *The Wire*. Then there is my own father, who uses his middle name, which he shares with an uncle who was a casualty of World War II, as well as a guy who died in a carjacking. In other words, you are not at all safe assuming that search results for your name are all about you.

- Your name, forum posts, and other information you reveal online can assist someone in discriminating against you and not getting caught. For instance, your last name may indicate your ethnic background, possibly including your country of origin and religion. Posts about your hobbies may indicate your political affiliation, social class, medical issues, and even your temperament. For instance, people will have vastly different expectations of someone who trains in combative martial arts, hunting, and weightlifting compared to someone who meditates heavily, enjoys video games, and loves to cook. All six of the items listed apply to me, but I would rather reveal them selectively to a potential employer. It's better to control this if possible.

One thing that makes all of this stuff worse is that information on the Internet is notoriously hard to get rid of. Search engines cache at least part of much of the information they come across, archive sites mirror existing sites, and sometimes people share their own downloaded copies of things that have been removed from the Internet. The Internet is forever, so if anything negative is out there, you're going to have to take steps to deal with it.

There are a few things you can do to counter the negative search results that appear when you search for your own name, few of which involve removing the offending content. Because it is so difficult to remove content from the Internet, other strategies are called for.

Finding Out What's Out There

First, however, you need to figure out what you are dealing with. You not only need to have a way to find negative things that are already out there, but you need to be able to quickly be alerted to new problems. While it's tempting to simply start trying to remediate the problems you are aware of, doing so before you are aware of all the problems is not a good idea. Unless you have just searched for your name in one of the major search engines, your perception of the biggest problems is unlikely to be accurate. Search engines prioritize things that you might not expect, and it is better to be prepared than surprised.

The most obvious thing you should do is put your name into a search engine and see what pops up. Don't do anything with the information you find yet, just take notes about what is being said and where it is being said. You'll also need to repeat this search in relation to where you've lived in the last decade or so. For instance, I might search for "William Gant near Nashville TN" and

"William Gant near McMinnville TN." You should do the same for major skills that you are listing on your resume. When doing this, take special note of the following:

- What was said, and how it was said. If someone is praising you in some way, take note of the way they expressed it. Similarly, if it is negative, do the same. Also make sure to get the usernames or forum handles that people are using.

- If the post was something you said, take note of the username that you used and the web site.

- If any links are included in the post, take note of these as well.

This will give you a rough idea of what people are saying, including what you said about yourself earlier. Now you should also do similar searches for images, using the same criteria that were listed previously. People often forget this step, but any photos or videos that you have out there will often expose more information than you might think. I'm relatively lucky in this regard because I've never liked having my picture taken, so there aren't very many of mine out there. However, my attitude is far from common. Looking at a friend who is a bit closer to normal, I can see several things that might give an inappropriate impression or reveal information that allows discrimination. For instance, the following things are trivially discernible from photos that are publicly available:

- Your race, marital status, a rough idea of your age, your religious background, and your gender. While many of these things are going to be obvious when you go in to an interview, their availability online can enable discrimination before the fact. If you are concerned about this, it might be a problem. Whether this is a concern or not is very much a personal matter.

- Your past history of things that others find negative. This can be anything from alcohol and drug use to hobbies that tend to be stereotyped. People often make shallow judgments of others based on their pasts, hobbies, and pastimes. Others may not be judgmental or even view things positively. For instance, drunken pictures from a pool party during college might have been fun at the time, but lend a decidedly unprofessional air when you are trying to get a job at large accounting firm.

- These images can also reveal your usernames on other sites if they were used for a profile picture. This can be tied to other posts.

Make a full list of the usernames you have, and do a search for that username on the broader web. While not all of the posts you find will necessarily be you, you need to be aware of any that can be misidentified. You should take note of all these that you can find.

Social Media

In addition to things found on the open Internet, you also need to take some time to deal with social media. Because people tend to approach social media differently than they do many other web sites, you need to take special steps to make sure that these sites don't give a bad impression. In this section we'll discuss a few sites and the problems that they can cause. Each of these sites requires a different strategy to manage. Any new social media sites will probably have similarities to one or more of the sites in the following list, and you can clean up your online persona in similar ways:

1. LinkedIn—This site is generally professional, so make sure that you have your profile and resume cleaned up (as suggested in Chapter 5). In addition, pay attention to any recommendations you've gotten and what they say about you. Obviously, if you have any unprofessional or inflammatory posts, you should probably remove them.

2. Facebook—By default, your Facebook profile could potentially have a lot of personal information on it. In addition, because the social network is not intended to be professional, it's fairly likely that at least some of your posts and pictures on the site are not the sort of thing that you want an interviewer to see.

3. Twitter—Because of the short posts on this site, many people use it extensively and for very short posts. These posts can trend toward the kind of things that make hiring managers cringe. You also have to be careful about who you follow and posts you like, because those may count against you.

4. Instagram—Billing itself as a photo sharing application, Instagram allows you to share photos with followers. Depending on what kind of photos you are sharing, this can reveal information that you don't want revealed, or even damage your job prospects. On the other hand, photography can also humanize you in a way that is helpful.

Generally speaking, with social media you want to make sure that a few things are true (and remain true). There aren't really many rules, but they are very important to protect your image. These are as follows:

1. Don't have any obviously offensive posts that are visible with a cursory search. While it's probably impossible to remove everything that could be a problem from social media, removing the things that are easiest to find is sensible.

2. Don't allow other people to post things on the publicly visible parts of your profile. Not only do you want to mitigate the damage you caused posting while not considering the future, but you need to be able to keep other people from doing the same and involving you in their mistakes.

3. Have a good separation between the professional parts of your persona and the personal ones, without those things crossing. This may mean hidden posts or even separate accounts, but the idea is to keep from leaking personal stuff into your professional profile. It has the added benefit of making things less annoying for people who have no interest in you professionally.

Continual Maintenance

In addition to cleaning up your online persona and protecting yourself on social media, you also want to be proactive in regard to your online persona. You probably also don't want to spend a ton of time on a regular basis searching your name in Google to see if anything bad has showed up. Thankfully, there are a few services you can use to monitor search results for your name. There are a number of both free and paid tools available that can help you with this process.

First among them is Google Alerts (www.google.com/alerts). With a Google account, you can set up alerts for new things that match the criteria that you specify. In this case, you would probably do well to set up alerts for anything matching your name and your current city of residence. You might also expand a little to include locations where you visit frequently. Google alerts will be emailed to you as the search finds new matching pages.

Second, if you are active on Twitter (or your social circle is), you should make use of Mention (https://tweetdeck.twitter.com/). This tool lets you monitor Twitter for conversations that involve you, which can be helpful not only for engaging with potential employers but also for monitoring your interactions on Twitter.

There are a lot of other tools available for managing particular aspects of your web persona, but Google Alerts and Mention will be enough to get you started. There is another thing you should do as well. You should set up a recurring reminder on whatever calendar software you use to manually review your search results. While the software helps, you should still check things manually in case something has changed.

Cleaning Up the Mess

Once you have a good idea of what kind of problems your online presence can present during an interview, you need to figure out how to deal with them. This isn't necessarily as simple as you would like. There are a number of reasons for this, chief among them that the Internet does a fair job of storing things, even if you (and the rest of society) want them gone. The bad news is that if you've done anything sufficiently bad that has gotten wide attention, it may well stay online forever.

That said, your life isn't over just because something bad is out there. There is a lot you can do to mitigate most things, and even the worst things can be drowned out by good things. Given the list of items you found earlier, you have a few options for cleaning them up.

If you made the post and still have access to the account used, the easiest thing to do is to simply delete the post and move on. This doesn't mean that it is gone, especially if it was a scenario where other people could quote you and reply, but that's probably as far as you can get.

If someone else has made the post and it is inaccurate, you can often contact the owner of the web site and ask for it to be removed. When posts are defamatory, or border on libel, it's often easier for the owner of the web site to remove it, rather than risk a legal battle. Additionally, you can also ask the person that made the post to remove it. This may or may not work. If the post is particularly defamatory, you might even want to consider legal action. However, bear in mind that legal action can also create public records, in addition to being a distraction.

In addition to removing things from the Web, you should also take some time and figure out how you want to deal with things that cannot be removed. If you think the interviewer has seen the posts or you are worried that they have, you should consider how you can explain the situation. Remember that the interviewer is human too, and they probably have things in their own past that they would rather not admit. If you can gracefully handle their questions and show what you learned from the situation, it can often give a more positive impression. If something negative is online and you can't remove it, you need to be working on a way to explain it to your benefit.

However, there are ways to make it less likely that negative things will be found. In the next section, we will discuss those.

Better Than Cleanup

While trying to clean up your online profile is a laudable goal, it will not fix everything. Thankfully, there are a few more things that you can leverage in your favor. Anyone looking into your online profile using a search engine is probably not going to be looking past the first few pages of results, simply because it is too time-consuming to do so. Therefore, if you can fill up the first few pages of search results with good things, you've effectively erased the negative things, at least as far as a cursory search can reveal. The added benefit of this approach is that it completely changes the nature of the discussion during the interview, by changing the interviewer's perception of you as being someone who "screwed up" into a perception of you as someone who is doing the "right things."

If you've taken the advice earlier in the previous chapter to start participating in forums, you probably already have a good start on fixing your search results. As long as the forums you participate in are positive places with good content and show up in search engines, the less likely it is for anything bad to show up in the first few pages of search results.

Additionally, if you blog regularly, especially under your own name, people will often visit your blog to learn about you, rather than focusing as much on search engine results. Search engine results do return a lot of false positives, and it can be time-consuming to sort them out. A good blog that documents your journey also does a lot to humanize you to the interviewer and make them invested in your story. In fact, it can often make people invested in you to the degree that they reach out to you about jobs.

In short, putting yourself out there and producing content will not only improve the impression you make online, but can be used to effectively change the tone of the conversation with an interviewer.

Summary

In this chapter, we discussed ways of cleaning up and adjusting your online persona in order to shape a prospective interviewer's perception. Many companies will search for you online before even extending the offer of an interview, so it's important. We discussed how you need to search for things that might be a problem, followed by how to correct the things that you can correct. We then followed that up by explaining how regular online participation in forums and blogging can make it more likely that someone finds positive information about you when they are considering hiring you.

Researching Before the Interview

Before interviewing, you need to spend some time researching the places where you will be applying. Besides preparing you for interview questions, such research can also help you identify places that aren't a good fit for you. There are a lot of ways to do this, and we'll be discussing these in this chapter.

While researching a potential employer will put you ahead of the crowd, in terms of preparing for the interview, that's not the only thing you need to be looking into. You also need to learn as much as you can about the industry, the competition (if any), and what kind of culture the company has.

Many developers don't put a lot of effort into researching potential employers, so even a little bit of work in this area can really pay off. In recent years, it's been pretty easy to get a software development job, especially once you have a little experience. As a result, it's common for developers to interview for a job while knowing next to nothing about a potential employer. The thought process is that one can always get another job if this one turns out to be terrible. At the time of writing this book, that is still largely true.

© William Gant 2019
W. Gant, *Surviving the Whiteboard Interview*,
https://doi.org/10.1007/978-1-4842-5007-5_8

However, this strategy will not work forever. There have been times when it was much harder to get a good software development job than it is now. Particularly after the market crashes of 2000 and 2008, there were several years where the jobs were scarce and the competition fierce. It took more than just skills and experience to get a good job, because there were more unemployed developers than jobs for them to fill. A lot of people I knew not only lost their jobs but were unemployed for a long time while the market recovered. Once they finally managed to get another job, their pay was often lower. Years later, some of them are still feeling the effects.

In many cases, a little research about the companies where they were interviewing would have made all the difference. While research alone won't make you into the best candidate in the world, it can easily make you look far superior to other people at your skill level who did not prepare.

This chapter is about how to research a company before applying for a job. This will help you formulate better answers to likely interview questions, help you gauge whether the company is a good fit for you, and even help you get some idea of what trade-offs are involved if you take the job.

Why You Need to Research Before Interviewing

You might wonder, as I did when I first started interviewing for software development jobs, why you need to do a bunch of research on the companies where you are interviewing. After all, aren't you going to learn on the job if you get hired? That's actually not an unreasonable question and the assertion that follows is more reasonable still.

However, "learning on the job" can be one of the most painful ways to learn something about a company. I'll use some personal examples to illustrate my point:

- One company where I worked was shedding customers and employees at an alarming rate while hiring less qualified staff (myself, at that point) to try to fill in the gaps. It didn't work and I lost my job during a recession.

- Another company was widely known to be a bit unethical, based on some of their interactions with other area businesses. I was fired for insubordination when I refused to do something illegal, and the place was raided by the FBI a few years later.

- Another company worked on neat technology, but had insane working hours. We're talking 80-hour weeks, with the expectation that you'd either get lunch from the gas station and eat it while walking back to the office or that you'd just skip it altogether and keep working.

- Another company was using mostly old technology, but offering jobs in new technology and then making people work on the old stuff after they started. This was on top of a commute through construction that was an hour each way. Plus they made me wear a suit every day.

One of the most critical things you can learn by researching a company beforehand is whether you actually want to work there. Employment is very much like marriage in that the right job is well worth the effort, while a terrible one is misery that you can't easily (or cheaply) escape. At best, most people will spend a third of their weekday working life at their jobs. When you include the commute, any overtime, and time spent at lunch, the number is well in excess of a third. Further, many people tend to take their work home with them, especially at an emotional level. A miserable job can make your entire life miserable.

Even mediocre jobs have consequences. They can often reduce your income over the course of your entire career by slowing down your acquisition of skills and experience, or simply boring you out of your mind. In addition, they may trap you in a technology stack or industry that is dying, meaning that you have fewer available opportunities over the long term. They may also force you to live somewhere you don't like, follow a career path that is unfulfilling, or just provide a constant, low-level stress. One might not think this is uncommon, but if you search for the word "unfulfilling" in Google search, the third suggestion in the dropdown is "unfulfilling job." It's more common than you want it to be.

In addition to filtering out unsuitable jobs, researching a potential job beforehand can make you more prepared for the interview. When you consider how poorly prepared many people are when interviewing, a little research can distinguish you from the crowd of people who "just want a job, any job." While the latter sort can often do a reasonable job, most hiring managers would prefer to hire people who seem to understand the company's goals.

There is a dynamic here that is useful to understand. It costs a lot of money to hire a new employee, whether it works out or not. From my own experience, the cost of hiring a new developer is easily in the tens of thousands of dollars by the time you include the cost of paperwork, equipment, training, background checks, employee time for conducting interviews, onboarding costs, and the time cost of other employees who are slowed down by a new coworker. It is not remotely close to cheap to hire a new person, and if they quit (or are let go), you get to do it all over again.

It's also not cheap to lose an employee, whether you lose them to a firing or to a resignation. This is particularly true of employees who are a bad cultural fit, unable to do the work, or who are simply difficult to work with. In many places, you can't simply fire an employee without a lot of preparation, which means that a toxic employee may well stick around for a while, damaging the morale of your existing employees and potentially introducing bugs and other problems that harm your clients.

There is also a cost to delaying a hire. An employer who is trying to hire more people is doing so for a reason. Usually that reason is that they don't have enough people to do all the work they have. When they wait to hire more people, one of three things can happen. The first is that they overload their existing staff, risking burnout (and the associated costs of staff loss). The second is that they have to reduce in the amount of work they take on, meaning that they miss out on opportunities. The third is that they take on more work than they can do, and then fail to deliver, which can cost them future opportunities and even existing clients.

Given these constraints, companies will often find themselves under different kinds of hiring pressures.

- A company that is really overloaded with work and growing quickly will often be under pressure to hire far more quickly. However, this is also true of companies that are losing employees quickly. Both are trying to avoid the opportunity cost of failing to hire.

- A company that is preparing for expansion in 6 months or more may not hire as quickly, because they would rather err on the side of adding personnel who work really well with the critical team members they already have. This can also be true of a company that can't really decide what they want. Both are trying to avoid the cost of hiring (and subsequently letting go of) the wrong person.

When preparing for the interview and when being interviewed, try to figure out what constraints apply for that employer. Sometimes it's obvious or they'll tell you outright, but you may need to ask. Not only can this help you determine whether you want to work there, but it will help you more accurately speak to the needs of the hiring manager. Compared to the cluelessness that many developers exhibit in regard to these things, your awareness of them will be a refreshing change.

Starting Points for Research

There are a few basic things you need to learn about the organization before interviewing. These things will often point toward other information that will be useful to gather. In addition, proper research will help you ask more intelligent questions during the interview. Not only will this potentially lead to a much better interview experience, but it will make the hiring manager remember you.

After this section, we'll discuss how to obtain this information, but for now, we'll discuss some of the things your research should help you learn. If you have the following information, your interview will be much easier. Essentially, what we are trying to do is get enough information that you are overprepared for the interview. This will make it much easier to stay calm, will give the interview a much better impression of you, and will let you be more productive in your new job more quickly.

Skills the Organization Needs

Probably the most important thing your research should uncover is a breakdown of what kinds of skills are in need at the organization. While you may have found a job listing that suits you, you may not have seen all the jobs that the company has available. You can learn a lot about what the company is doing based on their available job listings, including the technologies you'll be working with, where the company offices are located, and possibly what their future plans are.

When you start examining the other jobs the company has available, you can easily learn what technology they are using (and possibly what they plan to use). For instance, if you find that they are looking for both people to maintain legacy ASP.NET web forms applications and people with experience in using Node on Amazon's AWS infrastructure, you could reasonably assume that they are dropping ASP.NET for Node. This information could potentially be useful in several different ways.

- You might be an ASP.NET developer who has no intention of learning Node. If this is the case, you might want to find whether they will be keeping ASP.NET in their environment, and for how long. You might not want to take the job if it is going away, so this is something you'll want to learn in the interview.

- You might be an ASP.NET developer who wants to switch to Node. In this case, you might want to work for this company because it will give you an opportunity for a smooth transition.

- You might be interested in working remotely. A company that is starting to move toward the cloud might be a good place to do that, since companies that have all their infrastructure in-house are more reluctant to let you use a VPN.

In addition to learning about technology jobs that the company has available, it can be helpful to learn about what else the company is doing. For instance, if they are hiring a number of sales staff, it might indicate that their sales team is growing pretty rapidly. This often indicates that they may be looking to hire more developers soon as well, to handle the increased work that the sales team produces.

Similarly, you will probably want to check and see if any positions are open that would be above you in the chain of command. It doesn't necessarily mean that the environment will be changing for the worse, but new management can often have wildly different expectations and plans than what might otherwise be indicated by a job posting or online reviews.

Products or Services They Provide

You also need to spend some time learning about the products or services that the organization provides. Even governmental organizations have a purpose, and you need to have a good idea of what that is before you try to interview there. There are numerous advantages to be gained from knowing the purpose of the organization.

For instance, you might have some experience in the industry. If you know this before the interview, you might be able to talk to other people you know who are in the same line of business. This might even give you some ideas as to who in your circle of contacts might be an especially good reference. In addition, knowing some of the insider language of an industry makes it far easier to come across as knowledgeable.

Additionally, if a company offers a product, it can be really helpful to examine how that product is delivered (if digital) or the supply chain required to create the product (if it isn't). Even if you don't have experience in this particular industry, understanding how something is made and delivered can often mean that you have useful knowledge that you can leverage in a new job. There are decent odds that you have some connection to some part of the process. You should use this to your advantage as much as possible, and it is far easier to come up with ways to do that when you know in advance.

If you can use the product yourself, see a demo of it online, or otherwise find out more about it, you should do so. Not only will this give you a lot more information about the company but it will often give you a look at things like their design philosophy and the language they use in their marketing (more on this later), and even give you some insight into how their product competes with other products in their field.

Who Are Their Clients?

It's also a really good idea to learn as much as you can about the potential clients or consumers of the organization's products and services. While developers don't usually deal directly with clients and users, the feedback from clients and users is a huge driver of the kind of work that you'll be doing. In addition, if their clients have special constraints on their use of the product, that can be really helpful to know going in.

For instance, let's say the organization's clientele are residential property inspectors who use some software provided by the company to record inspection issues (and their severity) for the purpose of determining whether a home is a good purchase. There are several interesting things that this tells you about how the client will probably want to use your software. First, they are probably going to want touch-enabled devices with an onboard camera. Secondly, they are frequently going to be using these devices in situations where an Internet connection is extremely slow, if it is available at all. Therefore, they are likely to want to record everything on their device and then synchronize it at their office. Third, they are going to be moving around a lot, potentially within enclosed spaces that might not be particularly clean. Their worst-case use of the application might well be a house where someone has been hoarding junk, and they may well have to crawl through a nasty, unmaintained crawlspace while using your application to document issues.

This tells you a lot about the application. First, a standard web application simply will not work in this scenario, so it behooves you to think a bit about how they might structure the application. Secondly, the user interface for the application will probably be touch-based and will need reasonably large buttons to support that (if one assumes that they are probably wearing gloves when working with the tool). Additionally, they will be working in environments that are pretty hostile to electronics, so you can expect that the devices will be ruggedized, which can bring interesting problems into the mix. Further, because their clientele is probably sitting at a desktop computer when looking over the inspection reports, that means that the reports need to be synchronized with a server on a fairly regular basis.

In addition to being an impressive way to show insight into the organization's customers for the purposes of building an application, it can also win over the nontechnical interviewers in the room. In many organizations, at least some of the people who conduct developer interviews are either nontechnical or were technical so long ago that they aren't there for technical reasons. Sometimes these individuals are hard to win over for the average developer, because the average developer focuses on the technology, not the problems being solved by the technology. If you show that you understand the constraints that their clients are dealing with and how they use the software, these people may well become your strongest advocates. If you've combined that with properly preparing for technical questions, you are far more likely to get the job.

Their Competition

Along with learning about the clients of an organization, it's a good idea to learn about their competition. You can learn a lot of surprising things by researching competition, including what features the competing product has, how the competing product is sold, and how well the organization's product is doing when compared to the competition.

When comparing the competition to the organization where you are interviewing, you should look for a few things. First among them is the differences between the products or services and how they are marketed. For every difference you find, you should be asking why it is there. This is harder to do for services than it is for products, but it is still possible.

Differences in software products, web sites, and physical products often point toward different usage scenarios. Sometimes these differences will show that the organization is targeting slightly different audiences, while other times the differences will be so slight that they are clearly intended for the same audience. If it's the latter, be sure and take note of the competitor's information, as it's often useful to ask about anything interesting you see on a competitor's site when interviewing.

Their Business Sector

Additionally, you should find out a little more about the industry in general. Almost every industry has its own insider jargon, its own shows, its own industry-specific schedule, and varying degrees of expectation for their employees. For instance, accounting firms in the United States are often extremely busy during the spring before taxes are due, while tourism-related businesses are often busy during the summer months and around school breaks.

Nearly every industry has its own specific industry shows. Industry shows are a great way to find out about both competition and your target company's strategic partners. While it could really help you if you find some way to attend a conference or show, these tend to be a little too expensive for an individual to consider attending. However, businesses that attend a major conference will often produce a lot of marketing material as part of the effort. Because of the cost of creating this material, much of it will be reused in the company's marketing efforts. As a result, you can often find a lot of good information about your target organization's products and how they are sold on their own web site. You can also do the same for their strategic partners and competitors in order to build a clearer picture of the market forces that are creating a job opportunity for you. It's a lot easier to do well at interviewing when you know why you're really there.

Recent Organizational News

You should also look for recent news stories about your target organization. This is especially true of larger organizations, as positive news stories can often give you a little more information about how a company is doing financially. Furthermore, it will help you identify key players in the company, such as the CEO, CFO, CIO, and other officers. If the company is publicly traded, you should also take some time to look at their stock price over the last 12–18 months. While this may not tell you too much about how the company is really doing, it may give you some insight when combined with other things you observe.

Additionally, be on the lookout for negative news stories about the organization. You should be especially careful to note if they are involved in any major litigation, public scandals, or if they've had bad news about their stock prices. These things can possibly indicate an environment with some risk. While the risk could mean that the job isn't entirely stable, it can also mean that the job is an excellent opportunity for growth. In the chapter after next, we'll discuss how to determine the level of risk by asking strategic questions during the interview.

Be careful about how much trust you put in news stories about an organization, whether they are positive or negative. While public perception of media trustworthiness may or may not be on the decline (oddly enough, depending on which news organizations you trust), the trustworthiness of the media itself isn't the biggest issue you face. A large portion of the information about a company in the news is going to be fairly biased in one direction or the other. It's probably worth explaining a few things that might contribute to this phenomenon.

- Many "news articles" about companies are thinly disguised press releases. They will be overly favorable to a company, unless their public relations or marketing department is terrible. This doesn't mean that they don't contain useful information for a potential interviewer, just that you need to understand that everything will be spun in a way that makes the company sound good.

- Many other news articles that aren't obvious press releases are still heavily influenced by the company. It's a fairly common marketing tactic for a company to use sites like Haro.com (HARO is short for Help A Reporter Out) to exert influence on the media by acting as sources for reporters.

- If the article is the result of litigation or scandal, the various parties involved are more concerned with telling their version of events instead of the truth. It's risky to assume that anything you are being told is truth until it's been sorted out by the courts, the police, or whomever is handling the situation.

- Stock prices are also a poor way to tell what is going on. While you can generally expect that they are correct and that financial reporting is accurate, any speculation as to the reasons for the numbers is best taken as an opinion.

- Reporting about an industry is also of dubious value, especially if the subject matter is more technical. In such cases, the reporter is often relying on other people for explanations of how things work. Explanations given for the purposes of reporting may be of dubious quality, as they are designed to be compact rather than correct.

News stories about an organization, its products, board members, and competition can work together to give you a lot of good information that will help in an interview. Later in this chapter, we'll talk about some ways that you can set things up so that you passively receive information that would be useful while working at a particular organization. If you leave these in place after you get hired, they'll often help you considerably once you have a job as well.

Quality of Life

Regardless of how good a company's performance in the stock market is, you'll hate working there if the environment doesn't suit you. While we'll discuss some of the hygiene factors that one should inquire about when interviewing in a later chapter, at the moment, you really need to get a rough idea of how the company treats their employees.

If an organization treats their employees badly enough, you can easily find out from news stories about the company. While that's rare, it can be a really good indicator that you don't want to work there.

However, the "medium stress" level of company quality is a bit harder to discern from simple research. You may have to dig a bit, including asking other developers in the community and finding sites on the Web that give reviews of companies from the inside (we'll go over some of those shortly).

However, you should be aware of a lot of caveats when listening to what is essentially gossip about a company. If enough people have heard the same thing about a company, that only means that it has a higher probability of being true than if only one person is saying something about a company. It does not make it true. Never take any of this stuff at face value, because the majority of people who will take the time to say anything on a review site may be doing so simply because they are emotional about the issue at hand. It can highlight real issues that you need to investigate, but it does not constitute an investigation itself.

This may make it sound like you shouldn't listen to what is being said about a company before applying there. That's not the intent. Rather, you absolutely should be listening to the rumors about various companies in your area. Just be careful not to trust rumors too much unless you really trust the people who experienced the environment first hand.

Unique Selling Proposition

With all the previous stuff in mind, you should also figure out your target company's unique selling proposition (hereafter referred to as their USP). The unique selling proposition is essentially a sentence or two that describes what benefits a product or service offers that differentiate it from its competitors. The previous items we discussed should have given you enough information to figure this out.

This may sound like a marketing thing (and it absolutely is), but it's also important for you to figure out before an interview if you can. This will allow you to tailor your responses in the interview to the values the company has, as expressed in their unique selling proposition.

You'll probably notice I didn't suggest at any point that you examine an organization's mission statement to understand anything about them. That's because mission statements are not a really good proof of anything, other than proving that the company wrote a mission statement. A cynical friend of mine tells me that he doesn't trust what someone says their priorities are— he trusts what they spend money and time on as examples of what they prioritize. The unique selling proposition is the same principle applied to companies instead of individuals.

Regulatory Environment

In addition, you should take some time to research the various laws that an organization may have to operate within. This can be very instructive, as sometimes those laws create situations that drastically change the way you have to treat code and data. In addition, there may be a variety of standards that inform how their organization will operate, including industry standards. If you know these going in, you should casually mention them when it is appropriate to do so.

There are lots of laws on the books in every country, probably to the point that the definition of a country is a place with "more laws than you can ever read." Many of these laws touch on business conduct in various ways and greatly impact the way your day-to-day routine will go. Here are a few laws and standards that you'll run into, regardless of what industry you're working in:

- HIPAA—Covers US standards for patient privacy. This covers healthcare privacy for providers, health plans, and business associates of HIPAA-covered entities. Designed to avoid disclosure of some of the most sensitive data you have, it can also generate rather spectacular fines if you screw up.

- PCI—Not a law, but a set of standards governing the storage, processing, and transmittal of data related to credit card payments. If your company is engaged in e-commerce, some level of PCI compliance will almost certainly apply. These standards ensure that companies treat credit card information with the level of security that it needs.

- GDPR—The European general data protection regulation defines how businesses and public sector organizations can manage the data of their clients. While at the time of the writing of this book, many of the regulations have not yet been tested in court, GDPR promises substantial fines to ensure transparency, accuracy, confidentiality, and accountability for sensitive data stored for citizens of the European Union. This act constituted a sweeping change in the way private data is handled for EU citizens and will probably end up impacting US citizens as well eventually, simply because it's safer to get it right for everyone than to have two entirely different classes of users.

- CAN-SPAM—Sets the rules for commercial email in the United States. This keeps email marketers from using deceptive email marketing tactics, sending unwanted junk email, and defines rules for how opting out of email marketing should work.

- Sarbanes-Oxley—Establishes auditing and financial regulations for publicly traded companies. "Sarbox" as it is not-so-lovingly called requires a large number of accounting and system access controls designed to protect shareholders and the public from accounting errors and fraud.

- Section 508—Governs accessibility standards for technology used or maintained by federal employees and members of the public. These standards define what you can and cannot do with technology and the ways in which you have to accommodate individuals with disabilities.

You should take time to learn a little about the preceding standards, simply because they are extremely common in a variety of circumstances. Even if the organization you are researching is not subject to these standards, they have a way of cropping up anyway, as it is difficult to avoid sensitive data and governments in most industries. In addition to these, you should also examine what major regulations are in force in the industry you are targeting. You'll see a lot of things done simply to comply with regulation that might not make a lot of sense otherwise.

Organizational History

You should also be trying to find out a little about the organization's history. While you don't have to know everything about how the organization came to be, there are a lot of things that you can infer about an organization based on its history. The following items are generalities, but they are often true:

For instance, a brand new organization that is still small is probably pretty scrappy. Unless there is venture money (or better yet, government money) backing it, the pay and benefits may not be as good as a larger, more stable organization. However, the opportunities for advancement and personal growth may be better, simply because there are fewer people. In such organizations, it's common to learn how to do a variety of tasks, simply to help keep the whole thing running.

An older organization that has remained small is often a family business. If they've been around a long time, they may be very stable, but it may be more difficult to advance your career. On the other hand, such places are a good place to be when you are more in need of security than advancement.

An older, huge organization is often publicly held (or governmental). Such organizations may have fairly good benefits and a stable career path over the long term. They also tend to be heavily specialized, which means that your job description may be narrower than you'd like. Nevertheless, many of these organizations do an excellent job of training their people. They are also a good place to work when your life is changing a lot. It's much less stressful to get married, have children, deal with sick relatives, or go through other major life events while working for a larger company, as they often have more generous insurance and sick leave. They also tend to have enough people to actually do the work, so it reduces the stress a lot.

A newer, huge organization probably has big money behind it. Most companies don't reach a vast size in short order without a lot of investment behind them. There are the occasional companies that manage it because of good timing, but they are rare. While these jobs can be boring, they can also be excellent opportunities to work with technology at a scale that is rare at smaller companies.

Additionally, you may find interesting details in the history of the company that may give insight into management attitudes. While this is less true of larger companies, the history of a small company can explain some decisions they make. For instance, I've worked at several companies that started in the 1990s. They often have very unusual ideas about what kind of applications they should build vs. what might be best for their customers. Depending on management attitudes, you can often make a huge improvement to the company's market position if this is the case. Similarly, I've worked at newer companies that had venture funding and did some really odd things because they were anticipating a lot of growth in a short period of time. Some of those things resulted in massive layoffs when they didn't work.

Where to Start Researching

To get the information you need to really do well in an interview, you'll need to do some digging in a variety of locations online. The first time you go through the process of finding useful information before an interview, it will take a while, but you'll get better very quickly. You'll also probably find that some information is more or less useful to you personally after you've gone through a few interviews. Like with any research of this kind, the goal is to collect just the stuff you need to make a good impression while not wasting a lot of time.

You'll also find that one or two of the sources I'm about to list end up being your go-to sources for information about a company. Your preference will depend a lot on your personality and the kind of information you are seeking. You almost certainly don't need all of these.

Their Web Site

The first (and most obvious) place to look for information about a company is their web site. While you should remember that anything you see there is going to be slanted in favor of the company (if it isn't, then that's even worse), there is a lot of useful information there that might be of use.

In particular, you can often find a lot of information about a company's history and working environment from their web site. It's pretty common for marketing departments to also post their press releases on the company web site, and most companies will at least have an "About" page from which you can get a variety of information. From the corporate web site, you can often find out if the company is publicly traded, its size, the products and services they are offering, names of strategic partners, and maybe the names of some key players in the company. You can also usually find out about their social media accounts from the web site itself.

You may not be able to find out much about the software development staff directly from a company's web site, since many companies are (rightfully) paranoid that their competition might poach their employees.

If the organization is a nonprofit or governmental agency, their web sites often have a lot more information on them, especially in their "About" pages. Because these sorts of organizations exist to varying degrees at the whim of the public, they spend a lot more time explaining what they do. It's still a lot of the same information that you would get from a corporate web site, just much more thorough.

The organization's web site should be the first place you visit to learn about them. Not only will it help you learn a lot of the basic facts about a potential employer, but it will also give you a starting point for the rest of your research.

LinkedIn

You should also check out the organization's presence on LinkedIn. While their LinkedIn page is going to present a lot of the same information as their web site, it will also allow you to find out if you know anyone who works there, or if any of your acquaintances can introduce you to someone.

If you have a little more time, you might also want to see if you can create some connections on LinkedIn. This can work especially well if any of the development team have active blogs or regularly participate in user groups and conferences. In addition, you can view the profiles of developers at the company to get a good idea of the kind of technology the company uses.

Pay special attention to any contacts you have that have already left the organization. In addition to being more forthcoming about what the environment is really like there, their LinkedIn profiles are more likely to show what technology they used.

Social Media

Once you've stopped by the organization's web site, you should check out the rest of their social media profiles. While these social media profiles are often still used by marketing people, they can still be a useful source of information about the products or services that an organization provides. One of the most useful things that such sites provide is insight into what the consumers or users think.

While you may not want to bring up negative user feedback directly during your interview, it's often useful to know what people are saying. If you're aware of the kinds of feedback the organization is getting, you may find that you have skills that are particularly useful for solving that problem. In that case, you might want to highlight the skills and how they can help. For instance, if you are interviewing at a company whose customers complain that their user interface is terrible, it might be worthwhile to put a little more focus on those skills.

When interviewing, it's important to remember that you are being hired to solve a problem for the organization. If users have taken to social media to complain about a problem, that's often an excellent place to start. If you combine this with carefully tailoring your resume as we discussed previously, you are far more likely to get the job than someone who didn't prepare at all.

Glassdoor

In previous steps, we discussed where to learn general information about an organization and the types of technology in use. We also discussed some options for finding out what kinds of problems their users are experiencing. However, in addition to the problems that users have, you also need to know what kind of environment the employees are working in.

While you could start with asking former and current employees about the environment, your questions will be better if you start out by doing a little research online. Glassdoor.com is a good place to do this. In addition to being a job search site, it also hosts reviews of many companies, written by former employees. While such reviews are less likely to be written by happy employees, they may highlight some problems that you should explore further.

In particular, pay attention to reviews that talk about things like excessive hours, poor planning, and the like that indicate organizational-level problems, rather than interpersonal ones. That's not to say that you shouldn't pay attention to interpersonal issues, but the organizational ones are more likely to be accurate.

It's important to evaluate sources of information, especially when the other side can't be heard. Don't try to guess whether they are correct or not (unless they are obviously incorrect); instead you should assume that they **might be** correct and figure out how you can prove it one way or the other. Remember that the kind of person who goes to write a nasty review online may not be the most pleasant to work around either, and take that into account.

Mutual Connections

Once you've gathered enough information to ask intelligent questions, now is a good time to reach out to any acquaintances who know about the organization. Ideally, you don't want to do this until you know enough both to ask good questions and to evaluate the answers you get from your acquaintance. While people won't exactly lie to you, they may not be entirely truthful either.

As long as you are trying to get information from people about a workplace, you run the risk of getting inaccurate information. Remember that a bad workplace can really mess up your life, put a strain on your relationships, and damage your finances. You have to make sure that the information you're getting is accurate.

Also remember that people have various reasons for telling you things, many of which have nothing to do with the reason you asked them. A disgruntled former employee can tell you plenty about how awful a workplace is, while the current employees may tell you that the former employee was awful. It can be difficult to determine which group you'd agree with. It's easy to characterize things as being true or false, but what will make a job pleasant or miserable is your perception. In other words, it's not about which group is giving you the most accurate information, but about which things you can tolerate at work and which you cannot.

This is why I recommend such extensive research online before asking a person anything directly. What you learn with online research can be verified by asking a person. The reverse is often not true, as the person's experiences are going to be extremely subjective.

When you make contact with someone regarding a potential employer, you need to get a few questions answered. If their answers match what you've already learned online, that's fine. If they don't, you may want to ask a few more. Here are some things to figure out:

- How do they treat their employees? Are they treated with respect?

- How stressful is the work environment? Is crunch time or overtime a frequent aspect of the work environment?

- How are the people? Are they pleasant to be around? What about management?

- How is the physical work environment? What about traffic?

- How bad is the technical debt in their codebase?

- How "corporate" are they?

- How long do people usually stick around there?

- Do they promote from within?

These are good starter questions when trying to get information out of an acquaintance. Ask some of these questions and see what you think of the answers you get. Pay special attention to the things that bother you and don't bother your acquaintance. While the things that bother them may be a big deal, the things that are dysfunctional and ignored will often tell you more about the environment than anything else. For instance, your acquaintance may be really frustrated by the amount of technical debt that the company has (hint: most of them have a lot) while being strangely comfortable with employees getting drunk at lunch and coming back to write code. If you dig into the former statement, you won't get anything that you don't already know—most companies are buried in technical debt unless they are very new. However, if you ask about the latter statement, you might learn about immature management or constant low-level stress that makes the employees cope by drinking. I've seen this one personally, and it's ugly. I avoided working at a company like this after asking questions of acquaintances who had worked there. When I found out that this behavior was going on at lunch, a little digging revealed a whole host of reasons not to work there, most of which my acquaintance was inured to through long exposure.

Their Technology Stack

Earlier we discussed how important it is to learn about what skills a company needs. In addition to that, it's a good idea to find out as much as you can about what technology they are using. Not only do interesting technology choices represent a good way to differentiate yourself from the other people applying for the same job. If the technology stack isn't immediately obvious, there are a few different tools you can use to get some idea of what they may be running.

- BuiltWith (https://builtwith.com/)—BuiltWith is a web site that allows you to find out what tools were used to build a web site or web application. While you can't get everything with a free account, it does do a fairly good job of getting enough information for a start.

- GitHub (https://github.com/)—GitHub is a popular source control provider. Many developers not only have their own free GitHub accounts, but they also link to their accounts from LinkedIn. Given the LinkedIn profiles of developers in the organization, you can often follow links back to GitHub and from there determine what tools the team uses. Be sure and do a Google search for "github.com [Username]" as well, since you can also find out more about libraries they are using if they've filed any bugs.

- Stack Overflow—Stack Overflow is a question and answer web site for software developers. Developers can both ask questions and answer them on the site. Many developers link their Stack Overflow profiles to their resume, while many others tend to use the same usernames that they use everywhere else. This makes it relatively simple to find profiles of developers and see what technology they are asking about.

- Other Job Postings—Additionally, if you've managed to find a few other job postings from this organization, looking at these will often give you more indication of what technology the company is using.

- LinkedIn Profiles of Current and Former Employees—If you have any connections to people who have worked in the organization, you might do well to check out their LinkedIn profiles as well. In their resumes, you can not only find out what technology was used, but how it was used.

- StackShare—StackShare allows you to find out what software is used by a lot of larger companies. While you can get this information from a lot of other places, it is far more accessible here for some companies than it would be otherwise.

While the tools I listed may not show everything that is used inside an organization, you can often find out what tools an organization is using for the majority of their work, simply by doing a little digging. Developers tend to talk a lot about the tools they use at work and their experiences with those tools. They also tend to ask for help in fairly public forums and report bugs in ways that are easily tracked back to their employers. You can follow this same trail to get a good idea of what the company is using internally. This will help you tailor your resume appropriately as discussed earlier.

Your Interviewers

If you manage to find out who will be conducting your interview, it's a good idea to do some research on them, especially via social media. If the company isn't particularly forthcoming about what they are doing, you may well find that the social media accounts of their employees can paint a better picture for you. In addition to looking for information about the company, this is also an excellent opportunity to find things (and people) in common with your interviewers.

A word of caution   A little bit of this goes a long way. You can reasonably say that you looked up your interviewers on LinkedIn because you are terrible with faces. You probably are to some degree or another, and no one will fault you for looking someone up on LinkedIn just so you'll know them when you see them. However, the same will not be true if you follow them on Twitter, attempt to friend them on Facebook, comment on their YouTube videos, and send them an email. Less is definitely more here. Get just enough information to be able to start a conversation and then get out.

There is a simple reason why you are well served to learn a little about your interviewers. You might find that you have things in common, such as political viewpoints, hobbies, and just general views on life. You also might find that you clash on some things in a big way. While major viewpoint clashes can be a good reason not to take job working with someone, you might also learn about things that you can casually bring up that will help you. For instance, if you attended the same college as one of your interviewers, that can be a good thing to bring up.

Many influential people will tell you that one of the best things you can do is to get other people talking about themselves. Far from being seen as a distraction, asking someone a question about something that interests them will often make them feel more favorably toward you. For instance, if your interviewer is an avid skydiver, you might ask them about the coolest place they've ever been skydiving. This may sound like a mere, idle question, but it can give you a much greater insight into the person you're working with and their attitude.

While you are being interviewed for a job, you also need to remember that you are not only interviewing the employer but also your future coworkers, and deciding whether you want to work there or not. You're potentially going to be spending a lot of time with these people, and you really don't want the experience to be unpleasant.

You certainly can deal with a terrible coworker or two for a long time, perhaps even for years. However, that is not a good situation compared to dealing with people you actually like. I've worked in a few places where I didn't really like my coworkers very much, and while in some cases it paid better, it really was never worth the extra money.

Toward that end, you should be gathering some information on your interviews to try and answer the following questions for yourself:

- How do I think this person will act when they are sick, tired, stressed, angry, or frustrated?

- If I'm stuck at a lunch table with this person, am I going to be miserable?

- Do I think this person will pull their weight on a team, or will I be picking up their slack?

- Is there evidence of personal habits that might make this individual difficult to work with?

- Is this individual, based on their career history, still going to be there in 6 months?

Looking at the LinkedIn profiles (and blogs, if available) of your interviewers can give you a lot of really good insight about whether they will be a good coworker or not. This information should be used both to help filter out potential employers and to provide stuff to talk about during the interview to get more information.

In order to do this well, there are a few places to research the other person. These are, in descending order of accuracy, the following:

- Mutual acquaintances who know both of you not only have insight into the other person's personality quirks, but they also have insight into yours. Ask your acquaintance if you think the two of you would get along and what you might have in common. Don't dig too deeply though—never ask anything that you wouldn't want displayed on a billboard.

- Personal blogs, podcasts, and YouTube channels of the person in question. When someone is regularly creating content, you can quickly get a good sense of their values and worldview, simply by listening or reading. A good example of this is the very book you hold in your hand—you've probably made several completely accurate judgments of me during the course of reading this much of it. Personality leaks into creative pursuits regardless of how well you try to avoid it—this is especially true if the person is producing regular content like a blog or podcast. While the population of developers who have active blogs or podcasts is low, if you get lucky enough to find one, it can tell you a lot.

- LinkedIn is another good place to learn about interviewers. While it isn't perfectly accurate, since it is a social network, it still will contain a lot of information that you can use. Just remember the purpose of this network and assume that anything you read was edited to make a job seeker look good.

- Other social media platforms like Twitter, Facebook, and Instagram can also give you a little more information. In particular, if the person has an active Twitter account and uses it regularly, you can pick up useful details by reading their public tweets.

It takes time to research potential interviewers, but the effort is worth it once you see it in action. When you have something else to talk about other than just code, you'll often learn things that you would have missed entirely in a conversation that remained entirely technical.

Logistics

Additionally, you should spend some time figuring out if it is actually practical to work for a particular organization in terms of your schedule, the distance you'll need to commute, and what else is nearby. For instance, if you have young children who are in school, working in an office an hour away from home can make your life much more difficult. In addition to determining whether the job can work out for you, you also need to make sure that you can make it to the interview on time while looking presentable.

Assuming your interview is in person, you need to start planning for the day of the event. You need to use a web-based mapping application to estimate the amount of time it will take you to get to the location from wherever you happen to be. Be sure and try to get a reasonable estimate of how long it will take to get there at that time of day. You should also add some extra time (I usually go for at least 30 extra minutes) in case there are issues with traffic or parking. Also, if you can figure out where to park ahead of time, it can save you a lot of frustration. In addition, knowing where to park and how to get from there to the location of the interview can keep you from getting stuck outside when it's pouring down rain or hot.

In addition to the interview logistics, you need to look into a few things to determine whether the job will be a good fit or not. While you probably want a job that can pay the bills, you also don't want a job situation that makes everything else in your life harder. To start, you should consider a few quality-of-life factors. You'll probably eventually find some of your own to add to this list, or remove a few, depending on your life circumstances, but you should at least consider the factors in the next few paragraphs of this section.

First, you need to know where you will actually be working. While this is typically the same place that you interview, it's not always the same. I had one job where I interviewed 5 minutes from my house at a Mexican restaurant. However, the job itself was conducted in two other locations, one of which was over an hour away when there was no traffic. It ended up being more like an hour and a half to commute there every morning, unless I left very early.

Speaking of traffic, once you find out where you will be working, you need to figure out what the traffic looks like for both sides of your commute. There's nothing quite so disheartening as accepting a job that "should" only be about 20 minutes away from home, but where the traffic makes that more than an hour. It's better to find this stuff out before even interviewing, because if the traffic is bad enough, you might not want the job at all. Alternatively, you may be willing to ignore the nuisance of excessive traffic if there are other transport options or if they are willing to let you work remotely.

You should also look into the available food options near your work location. This may not seem like a big deal if you pack your lunch most days, but having some good food options near work can be very helpful at times. I've worked in places where the nearby food options were all fast food or really expensive restaurants with slow service. You do need to make sure that you can have something reasonably healthy to eat if you do decide to go out for lunch. This can be even trickier if you have more stringent diet restrictions (including religiously based ones), allergies, or are a little more of a picky eater. If you do go out to eat regularly rather than bringing your lunch, this becomes even more important, as it can really have an impact on your wallet and your finances.

If you are going to be working in a busy downtown area and driving there, you also need to find out about parking options. Do so with an eye toward both your walking distance to your office and the price of a parking pass. You'll find that some employers cover parking and others do not. Some will also cover the cost of mass transit for you as well. While you'll probably have to wait until the interview itself to find out, it's worth noting before you sign up. If you recall the job I mentioned earlier that was an hour and a half away from my house to their main location, you probably won't be surprised to note that their other location was in a busy part of town where I had to walk over a mile one way in the hot Tennessee summer sun, wearing a suit and carrying a laptop. You don't want to find this out the hard way.

You should also consider what else is in the area near the job. For instance, if you have children young enough to need daycare, what are the daycare options nearby? Sometimes they may be better than what you can get at home. Some companies cover daycare as well, but it's not common in the United States. Additionally, you may want to consider other nearby facilities, such as nearby schools, malls, and concert venues that may make your traffic worse at certain times of the year.

Essentially, there are a lot of factors that you should examine before deciding that a particular job would be a good choice. Sometimes the quality-of-life factors that you don't consider can ruin an otherwise good job.

Summary

In this chapter, we discussed some research you need to be doing when looking for a job. We discussed specific information to examine to learn more about the working environment as well as improve your chances of getting a job there. Next, we discussed specific locations to look for information and what you will find there. We also discussed how to find out more about the organization's technology stack, which is critical for understanding your role there. After that, we discussed how to dig around for information about your interviewers without looking like a stalker. Finally, we discussed how to research the logistics of the job to determine whether it will be a good fit for you.

How to Get Through the Interview

Now we've come to the most critical part. This is the reason you bought this book, and it is probably the thing of which you are most afraid. This is what you will practice, and this is what I intend for you to absolutely crush when this is over. Interviewing well requires a lot of things to go right, and we're going to go through them in this chapter.

However, I'd be remiss if I didn't reiterate the importance of practicing extensively before you get here. If you are just reading this chapter when you are sitting in the parking lot before an interview, you're doing it wrong. I want you to internalize what is in this chapter before well you interview. You shouldn't need it when you are in the parking lot.

It's completely normal to be anxious about a development interview, especially if you've never had the experience before. That's one of the reasons you should practice regularly beforehand. It feels completely differently to go through something when you have no idea of what can go wrong. It's a lot easier if you've practiced beforehand, and it is easier still if you have regularly practiced your responses to the things that can happen.

© William Gant 2019

W. Gant, *Surviving the Whiteboard Interview*,
https://doi.org/10.1007/978-1-4842-5007-5_9

As mentioned before, whiteboard interviews combine the worst aspects of public speaking, writing code, and engaging in small talk with random strangers. Oh, and if you don't hit it off with the random strangers, you might have a little trouble paying your rent next month. It's very easy for nervousness to strike, causing you to make mistakes, which makes you more nervous, which causes more mistakes. The downward spiral can continue until it ruins the whole thing.

On the other hand, an interview going well will also result in a feedback loop, where it continually gets better as you do things right and everyone realizes it. The nice thing about this is that it doesn't take a lot to shift things in that direction. Assuming you can handle the technical aspects, a few other things that show that you'll be a good coworker will go a long way here. Writing in a legible fashion is a good example of that. There are a lot of little subtle things like this that you can do that show that you would be a good employee, without making it too obvious that you are trying to game the system.

Your Nerves Work Against You

The most important test to pass, besides the obvious coding test, is the personality test. In general, the personality test is implicit and is passed or failed based on how you conduct yourself. This is true of everything from the moment you walk in the door to the moment you leave the building. There is no seam between the technical aspects of the interview and the personality test; no matter what you do, you are dropping hints about your personality that will be interpreted.

While you should be positive during the process, you have to be careful not to come across as being fake. This can be challenging, especially if you are nervous. Try to think of the test as a challenge that you look forward to overcoming, rather than as something that will keep you from getting a job. While both are true, the former is more helpful than the latter.

This was an example of "framing." Framing is essentially a way of highlighting some aspects of a situation in such a way that you shift the interpretation of a situation. Being able to do this to yourself before going into a difficult situation is one of the keys to getting through tough things.

Framing also works on other people. For instance, there isn't a lot of difference between someone being "diligent" and someone being "anal-retentive," but you can bet that the two things are interpreted differently. The difference is largely a matter of perception (and social skills, frankly).

Now, framing might come across as if I'm encouraging you to be deceptive. I'm not. What I am telling you to do is to stop expressing things in a way that is arbitrarily damaging to your confidence and does nothing but harm you (note the framing in the previous statement). The fact is that when one is trying to determine whether someone is "diligent" or "anal-retentive," the difference is largely subjective. One person's "diligent" is another person's "anal-retentive." You should be honest, but there is also no upside in giving someone reasons not to hire you, especially when those reasons are highly subjective. For instance, it is better to say that you are "self-directed and motivated" instead of saying that "you do things your own way." While those two things are functionally equivalent in reality for many people, the former has a positive spin that will lead to a better impression.

The Personality Test

As someone who has given dozens of interviews and coached a couple of hundred people on the subject at varying degrees of intensity, I can tell you that the biggest mistake you can make (besides the obvious one of not being able to solve the problem) is to think that the problem is the only thing that matters.

When looking at an individual's performance, there are several key questions I'm trying to answer. Here are several of the most important:

- Can they work on a team?
- Can they work on my team with my current staff?
- Are they capable of saying "I don't know" or will they try to weasel out of admitting it?
- Can they accept correction when they are wrong?
- Can they defend their decisions without being defensive?
- How much handholding does this person require?
- Am I going to be telling them stuff over and over again because they don't take good notes?
- Are they capable of asking relevant questions?
- Does what I'm seeing match their resume?

The kind of questions you see here are the sort that are going through the mind of a hiring manager when you are standing up at the whiteboard, trying to remember how to do a bubble sort. And they almost certainly WILL have an answer for those questions by the time you're done. Let's talk through some of these questions in more depth.

Can They Work on a Team?

Software development is not a solitary activity and is unlikely to ever be. Due to this, employers not only want to know whether you can do the job, but they also want some idea of how you will interact with the team. After all if you can't work well with others, they are better off without you around.

Your ability to work well with others is first established in your interactions with others from the time you walk in the door. If you are rude to the secretary, it will get back to the interviewer. If you evade questions, get defensive, or are crass, you will make an impression that you don't want. This is true regardless of whether you are right or not.

Can They Work on My Team with My Current Staff?

Another thing that hiring managers have to worry about is the possibility of personality conflicts. It's entirely possible to have two people who are very productive on their own, but can't get along at all when they work together. This can be from a variety of sources, such as incompatible personal backgrounds, personal values, work habits, or even something as trivial as liking different sports teams. Sometimes people learn to dislike each other over a period of time, and sometimes they develop an instant dislike for one another. A hiring manager is going to be worried about this and will look for signs that it is a problem.

Additionally, people with strong personalities can often have arguments that don't bother either of them, but result in people around them being uncomfortable. For instance, two developers I worked with at a previous job would have loud, agitated discussions about technical minutiae. They were having a blast, but it made the people around them extremely nervous.

Similarly, your work habits may not be compatible with your coworkers. For instance, if you like to write tests as you are writing code and your coworker has a strong opinion that doing so is a waste of time, that's a potential flashpoint for an argument. There are a hundred things you can do that might set someone off, but if they come up during the interview, you probably won't get a chance.

While you may have strong opinions about a variety of things, the interview is not always the best place to bring them up.

Are They Capable of Saying "I Don't Know"?

Most of us have worked with someone who was a know-it-all. As someone who has hired a number of developers, one thing I always try to reveal in an interview is how a person reacts when they don't know something. Developers who make stuff up rather than admitting that they have a gap in their knowledge are extremely risky hires.

The fact is no one is expected to know everything, especially in an industry where everything is constantly changing at the rate of most software development platforms. Someone who tries to come across as knowing everything often poses significant risks to a company, whether it is from the amount of time they flounder around before asking for help, their tendency to misrepresent things to management, or their tendency to create interpersonal problems with other developers. Don't be this person. If you don't know something, you are better off admitting it and asking questions than you are guessing an answer.

Can They Accept Correction?

In the same vein, many companies try to determine how someone will react if they are corrected. We all make mistakes of varying severity at work, but a lot of people really handle it poorly. In an environment where things are changing rapidly or where people are expected to learn as they go, there is a reasonable expectation of making mistakes and learning from them. However, when one becomes too emotionally attached to their code, it is easy to see a correction as a personal insult and react accordingly.

Most developers with a few years' experience have dealt with one or more coworkers who are ego-driven and won't admit mistakes. Not only does this create a lot of morale problems, but it can also cause some of the better developers to leave. It's very expensive to have a developer on your team who won't admit their mistakes, as they have a stunning tendency to double down on their errors. It will almost certainly take more resources to clean up their mess than it took to create it, especially if they've already destroyed team morale.

As a result, a lot of development managers will try to correct you, simply to gauge your reaction. This doesn't mean that you have to immediately agree with what the interviewer says, but it does mean that you should answer in a reasonable fashion. This is a good opportunity for a dialog with the hiring manager, and you should take advantage of it.

Can They Defend Their Decisions?

At some point during your interview, a smart interviewer will probably start asking you questions about why you approached something a certain way. You need to be capable of explaining your answer and the reasoning behind it. If an interviewer is asking you to explain your answer, it doesn't mean that you did poorly. Rather, look at it as an opportunity to show your ability to reason about a problem and the potential implications of your solution.

How Much Handholding Do They Need?

Another thing that hiring managers must consider is the impact of a new person on their team. Any new team member is going to cause at least some disruption to the team, simply due to the effort required to bring them on board. A new developer will need passwords, a tour of the source code, a discussion of standards and procedures, and possibly some help with things they haven't seen before. However, some developers really need a lot of help for a long time to actually be useful on a team. In the meantime, they may well be a net negative for productivity for the rest of the group.

While a hiring manager may not be directly evaluating your ability to quickly become productive, they will notice if you seem like you are going to need a lot of help. While a certain amount of help is required for all developers of all levels, you want to make sure that you don't give the impression that you need an excessive amount of help to be productive.

Do They Take Good Notes?

Along with evaluating how much help you'll need to be able to be productive, hiring managers are going to want to see that you can retain the information you are given. While you may have excellent memory most of the time, the stress of an interview can make it easy to forget things. Write things down, especially if you will need to refer to the information later in the interview. Not only does taking notes help protect you from forgetting things, but the very act of taking notes may help you recall them later.

One thing that is particularly annoying in a new hire is having to tell them the same things over and over again. Not only does it waste another developer's time (or worse, the entire team's time if they repeatedly ask the same questions in a meeting), but a developer who constantly needs this kind of help will often forget other things they need as well. This can result in "interesting" coding decisions that cost a lot of time and money to clean up. In other words, you would do well to take good notes during interviews, simply to show that you have good note-taking habits.

Do They Ask Good Questions?

In addition to taking good notes, a hiring manager also wants to see that you can ask relevant questions. Many interviewers will ask you questions while leaving out important information that is required for an accurate answer. The goal is to see whether you ask reasonable questions while formulating an answer or make assumptions. Even though your assumptions may be right, it's always worthwhile to check them. We'll expand on this later in this chapter where we talk about the whiteboard, but you should get comfortable asking relevant questions about programming problems. It will take some practice to get used to coming up with good questions, but being able to do so is very helpful for your career in general.

Is Their Resume Accurate?

This may come as a surprise to you, but some people who apply for high-paying jobs lie, or at the very least, misrepresent their experience on their resume. This can be anything from someone who thinks they can get a senior developer job because they've installed WordPress once to someone who inflates their years of experience. If you interview a lot of people, you'll eventually realize that you need to be able to check the accuracy of their resume.

Toward this end, interviewers may ask you questions about what you did at previous jobs (if any) or what technologies you've used in the past. This will often trend toward asking detailed questions about how you used various pieces of technology, what you liked or disliked about them, and whether you would use the same tools again. While your answers may be interesting in their own right, they also show very clearly whether you actually understand the technology you used. This is extremely hard to fake, so if you don't know something well, it's best to be honest.

What Are Their Weak Spots?

In addition to figuring out what your capabilities really are, an interviewer needs to know about your weak spots. That doesn't mean that you should tell them directly. However, they will be asking questions to try and figure that out. Everyone has weak spots, whether they are personal weaknesses, things that you don't understand well, or even personality quirks. We all do. For instance, I'm terrible at making user interfaces look like something you'd want to use. It's just not a skill I really have. I can, however, make pretty decent software mockups and do a good job of figuring out how a user should interact with a system. But you don't want me doing artwork or choosing a color scheme.

If you look at what I did in the previous paragraph, it's a good example of how to own up to your weaknesses and redirect the conversation to a discussion of your strengths. It's fine to admit that you don't have skills (or interests) in certain areas. First, you really don't want to be hired for a position that requires a lot of stuff that you don't enjoy doing. Second, this is the time to have these discussions, because this also allows the hiring manager to consider where you might fit in the organization if they need you. While job requirements are certainly listed for a position, you'll find that sometimes the employer is willing to tweak the requirements of your job to suit your strengths and weakness. Thus, your approach should be honest and should allow them to find a space for you, provided that you are a good fit for their organization.

If you are directly asked questions about your weak spots, don't feel shy about owning up to them.

Body Language Hacks

In addition to a positive attitude, you'll also need to master a few tricks of body language. If you are like many people trying to get into software development, it's entirely possible that socializing and other interpersonal interactions make you nervous. That's completely fine. I used to be painfully shy myself, and while the discussion of how to work through that is beyond the scope of this book, there are a few tricks you can use to look more confident. And, if you practice them enough, as I did, you'll eventually look around one day and find that you have a level of confidence that you would never have expected. Handling interpersonal situations well creates confidence, which improves the ability to handle interpersonal situations. Thus, to some degree at least, you can actually "fake it until you make it."

Perhaps the most important thing (and the most difficult) for showing confidence is eye contact. This can be especially tricky to do well, because if you look away too quickly, you look shy or embarrassed. However, if you stare for too long, you'll come across as creepy. First, let's talk about how to "look someone in the eye" without being intimidated. The trick is you don't. Instead, you can look between their eyes. Most people can't tell the difference anyway, especially at normal interpersonal conversation distances. Just don't do it for too long (no more than a couple of breaths). I suggest either practicing this with close friends or random strangers somewhere where no one knows you and where you plan not to return. The idea is to get comfortable with people looking back at you.

The next critical thing to get right is your posture. You can certainly get a job, even with terrible posture that exudes a complete lack of confidence. It just makes it a lot harder.

To fix this, the obvious rules apply. Don't slouch and don't look at your feet. Keep your shoulders back. Basically, do all the stuff you were nagged about by adults when you were growing up. This is easier said than done, and you'll want to make sure that your whiteboarding study partner stays on your case about it when practicing. It can be challenging to get this right and isn't 100% necessary in any case, but it's worth practicing, especially if you have particularly bad posture.

In addition to posture, pay attention to how you react to being nervous. Your whiteboarding partner will help here if you don't already know. Nearly everyone has something that they do when they are nervous, whether it is biting their fingernails, playing with writing implements, or tapping their foot. You should at least be peripherally aware of your own nervous tics, even if you can't entirely suppress them. Don't be like the nervous kid I interviewed at a previous employer, who was playing with a whiteboard marker during the entire interview. At one point, he got so wound up that the marker actually flew out of his hand and nearly hit one of the other interviewers in the face. As the technical interviewer, I thought his skills were quite good, but all the nontechnical manager saw was the nervousness. He didn't get the job.

While you can't completely control them, being aware of nervous tics can help you mitigate the problems they cause. Once you notice yourself doing that, start taking deeper, slower breaths and remind yourself that you are there evaluating the company, not the other way around. Yes, they still have control over whether you get a job, but you need to calm down and continually remind yourself that they don't have all the power in the situation. This is something that you should be doing anyway, but a nervous tic (especially one that you notice) is a great prompt to remind yourself of that fact. If you manage to train yourself such that the nervous tic itself is a prompt to act more calmly, you'll really come out ahead. Whatever you do, don't get wound up on suppressing the nervousness; that will only make you more nervous. If you instead use it as a prompt to remind you of how much power you have in the situation, it will often go away on its own, especially if you also use it as a prompt to remind you to take slower, deeper breaths.

At the Board

OK, so we've talked about some general things that will make the entire interview easier. That was necessary, because there is really no point in doing well on the whiteboard portion of the interview if you leave a bad impression before you even get there. Bad impressions are harder to overcome than avoid. With that out of the way, let's talk about some things you absolutely should do in order to make a good impression when working through the actual whiteboard problems.

Write Down the Problem

When the interviewer is describing the problem to you, if it isn't already written down on the board, write it there clearly yourself. This does several things for you, all of which are huge improvements to the way that most people handle whiteboard problems.

- First, it gives you more time to think while not making you look like you are stuck. I don't know anyone, including myself, who instantly throws out perfect, eloquent solutions to a problem as soon as it is given, 100% of the time. I know a great many, however, who can do so 90% of the time or more, given just a little more time to think.

- Second, it forces you to restate the problem that the interviewer is giving you. This is a subtle way to make sure you understand their directions and to validate that the problem is appropriate. A lot of interviewers come up with problems on the fly, and sometimes those problems are nasty. I've seen problems given to junior developers that I couldn't solve if I had to. Restating the problem in one's own words often makes that obvious without being confrontational. This is not a small thing—I've given problems that seemed easy, but ended up being incredibly nasty more than once without intending to do so. If a guy writing a book on whiteboard interviews can mess up like this, it's reasonable to think that the same would be true of the average interviewer as well.

- Third, it shows the interviewer that you take good notes. It may not occur to them directly in this fashion, but doing this well can show the interviewer that you tend to only have to be told something once. Almost all development managers and senior developers out there have worked with junior developers (or even seniors…) to whom they had to continually reiterate things. This is a subtle way to indicate that you are probably not a member of that group.

- Fourth, it serves as a reference while you are working through the problem. You'd be shocked at how many people get about halfway through a whiteboard problem and promptly forget what they were doing. I've done it myself and you will too. Writing the problem down is cheap insurance against this tendency.

Make the Interviewer Participate

This is going to sound weird and there is nothing I can do about it but explain, but you also need to involve the interviewer in your solving of the whiteboard problem. There are a number of sound psychological reasons for this. First, it keeps them engaged in the conversation, which cuts down on the awkwardness of the interview. Second, people tend to like you more when they've done something for you, rather than the other way around. Third, it allows you to have interesting discussions and show a much better view of your skill level than you can possibly show by silent completion of a whiteboard problem. Remember that the whiteboard is used to evaluate your skill level, but it's not the only method of evaluation, even while you are working on the problem itself.

I know the second point probably sounded a little backward. "We like people more when we help them? Isn't it the other way around?" I suspect these are questions you are asking and they are entirely reasonable. Here's the deal. You are eliciting buy-in. They know they can work with you to get to a good result, because this literally gets them involved in doing just that. It builds rapport with the interviewer to a degree that you simply can't get any other way. If the interviewer is a terrible person to work with, this exercise has the added benefit of making that obvious very early on in the process, so that you can decline the offer.

So, now that I've (hopefully) convinced you to involve the interviewer by asking questions and discussing the problem, you are probably thinking about how to approach it. Here are some ideas that might be interesting fodder for the discussion:

- **How much data will I be processing?** Interviewers love to give string manipulation questions on the whiteboard, but the correct answers will vary a lot depending on how much data you are processing. Manipulating 10 kilobytes is a very different proposition than manipulating 10 gigabytes.

- **What libraries can I use?** Most development shops use extra libraries on top of the basic ones available in their language of choice. If you already know what those are, it's good to ask what you are allowed to use. If the interviewer is using other libraries for this sort of stuff and you are allowed to use them on the whiteboard, being able to do so with their library of choice is a benefit.

- **What kind of coding standards are expected?** Some places have weird quirks to the way they write code, and asking about these can yield interesting conversations. There are things I've learned from jobs I didn't get, and this is one way that I've done so.

There is one more reason you should actively involve the interviewer in your whiteboarding exercise. Some interviewers will fail to include critical pieces of information that you need to solve a problem. This may initially sound like a bit of a "gotcha" tactic, but there are several reasons that it happens. Some interviewers want to see how you handle ambiguity. Other interviewers are just distracted and forget things. Either way, you don't want to get tripped up on this, so it's wise to keep the conversation going.

Accepting Feedback

As you are working through the whiteboard problem, the interviewer may point out mistakes or ask questions about what you are doing. They may also prompt you to do certain things while building the code. Be reasonable about accepting the feedback. This is not a good place to get defensive, even if you feel like the question they gave you was a trick question.

Here's the deal. While you might be annoyed that they are interrupting your efforts to solve the problem, realize that you are in a much better position when they are interacting with you, rather than just simply watching. Now, instead of thinking that you are an idiot because you do something they don't like, you can have a dialog with them about your decision process. That's really a great position to be in, both because it's an opportunity to learn and because talking through an issue gives them a much better view of what it will be like to work with you. It will also show you what working with them is like, which can be quite valuable.

Finally, you may also be asked to refactor your code once you have completed the problem. This is also a good thing, as it both allows you to go back through the code and show that you are capable of cleaning up bad code, even if it is your own. Most companies have huge codebases by the time you get there, with code of varying levels of quality throughout. A good-sized chunk of code in almost any codebase is less than optimal or needs to be altered to better handle present conditions. Your ability to clean up code that you just wrote and to which you may have grown a little attached is something else that can differentiate you from other candidates for the job.

Now, you are probably also wondering what to do when you can't think of a way to make the code any cleaner than the way you wrote it. That's OK. Ask for a hint, and point out that you usually do a better job of refactoring and code cleanup a while after you have written it. That's almost always true of most developers, including those who are interviewing you. It's still a win, because it shows that you are open to input.

Overcoming Interview Challenges

Earlier, we talked about some things that can make interviews more difficult. It's time we revisited them with an eye toward turning them to your advantage. Companies that have problematic interviewing processes are not necessarily bad places to work, provided that you can succeed at the interview well enough to get hired. In the previous section, we outlined three situations that make interviews a bit more difficult for everyone involved. While these situations can make things more difficult, they also represent an opportunity to stand out from the crowd.

Whenever you run across a difficult situation in an interview, you should be trying to figure out how to turn it to your advantage. More than likely, other interview candidates for the same position are experiencing the same problems, and many will be woefully unprepared for them. With some preparation and the right attitude, such problems can be overcome in a way that makes you look better than the other candidates.

Remote Interviews

Interviews and live coding exercises conducted remotely are more difficult to evaluate. There are several reasons for this. The first is that technology will often cause problems in remote interviews. Whether it is an Internet outage, sound/video issues, or just the frustration of getting everyone into a meeting at the same time, technology problems are a constant source of irritation on a remote interview. In addition, remote interviews also make it less likely that either party will even see much of the nonverbal communication that occurs during an interview. These problems can easily result in misinterpretation, completely missing critical information, and being hard to understand.

There are a few things you can do to mitigate this, both before the fact and during the interview itself. If you find out you are being remotely interviewed (or given a coding test), you'll have some time to prepare beforehand. You should start by making sure that you have the latest version of whatever software will be used for the remote call, and that you have a rough idea of how it works. You can usually find the latter out on the software company's web site. If the company doesn't specify the software required, it's a good idea to contact them to find out.

Besides the software being used in the remote interview, you'll need a few things to do a remote interview well. These are as follows:

1. You need a computer microphone of reasonable quality. While you don't need a studio-grade computer microphone, you do need to verify that your microphone works and produces audio of a quality that allows you to be understood. It's worth testing this out with a friend if you are unsure.

2. You probably should also get a decent webcam for your computer. Again, this doesn't have to be top of the line, but it needs to work well enough that things like your facial expressions are visible when you are on a call.

3. You need a quiet place for the interview to occur. While you definitely can interview in an environment with a lot of background noise, it makes the whole process more difficult and increases the odds of making mistakes. If there are a variety of noises in your environment, you might want to consider getting a gaming headset for interviewing, because they do a "good enough" job of cancelling out background noise and are widely available. Reasonably good gaming headsets can be had relatively cheaply and can help reduce the amount of noise that is picked up. They also can make it easier for you to hear what is being said, without being distracted by background noise.

4. You need an Internet connection that is fast enough to handle video and audio without glitching. Depending on where you live, this may be more or less difficult to accomplish. If you aren't able to get a fast connection where you live, you may need to see if you can use a friend's connection for the purpose of the interview. You might also look into things like library meeting rooms and the like. Sometimes these are available for free or for a reasonable price. Regardless, make sure the connection from your machine to the open Internet is fast enough for video. This may require you to use a cable to directly connect to your router, instead of using a wireless signal. This can be annoying, but a slow connection makes your video choppy, your audio spotty, and generally annoys the people on the other end.

5. You need a computer that is decent enough to handle the remote interview or coding challenge. This usually won't require more than a normal consumer-grade laptop, unless they are expecting to watch you write code on your own machine.

6. You **DON'T** need your machine to sabotage your chances at a job. As a result, you need to make sure that nothing compromising is sitting out in the open on the machine itself. This can be anything from your desktop background, to your web browsing history, to random messages from your friends that look unprofessional. You need to make sure that none of this stuff shows up while a potential employer is looking at your screen. This may mean that you need to turn off your chat programs and other notifications before interviewing. You should also clear your browser history beforehand, just in case.

If you prepare well enough, your experience in a remote interview will be better than most of the people who go through such an interview process. Until you've interviewed enough people, it's hard to comprehend just how little the average person prepares in advance, especially with remote interviews.

While preparation will put you ahead of many other interviewees, you also need to do a few things to make sure that your remote interview goes well. These are as follows:

1. Sign in to the interview software early. Use the time before the interview to make sure that your microphone and video camera are working correctly; make sure your notifications are off and that you are ready for the interview itself.

2. Make sure that you've removed as many interruptions as possible before they occur. At a minimum, make sure you have some water to drink, have used the restroom beforehand, and have set your cellphone to silent.

3. Once everyone is online, make sure that they can hear you (and see you if it is a video call). This is common courtesy when doing these sorts of calls, but is easy to forget.

Cultural Barriers

Cultural differences can sometimes add difficulty to an interview, even if the interview is conducted in person. While a company should (and is usually legally required to) avoid bias against interview candidates, it's truly difficult to completely eliminate human bias from any process that involves human beings. I think we'd all love for that to be different, but it doesn't seem to be going away any time soon.

However, there is another way to look at this, and that is that some bias can often be overcome with a little bit of preparation. While you are unlikely ever to be able to completely integrate into someone else's culture, doing so is unnecessary. Instead, you only need to really be able to show that you can work well with their team in spite of any differences between cultures. That's a much easier requirement to meet, provided that you are careful.

Cultural differences are a minefield for both sides in an interview. It's also really hard to tell what cultural differences might be an issue and which don't matter. You might, for instance, see two people who get along really well with people from the other side of the globe who have entirely different religions and cultural beliefs. You might also see people who disagree on cultural issues that seem so small to you that you aren't entirely sure of the difference. In other words, it's nearly impossible to effectively mitigate the problems that arise from cultural differences in any consistent matter.

However, that's not the only option you have. Instead, your goal should be to accentuate the things that you (and by extension, your culture) have in common with the interviewer and their culture. Your goal should be to take the conversation away from how you differ from the interviewer and move it toward the similarities in your worldviews. That doesn't mean that you are being deceptive or being untrue to your roots, rather it's just important to recognize that you probably have more in common with the interviewer than you (or they) think and that it is more useful to focus on those things.

Disagreeing Interviewers

Another thing that can derail an interview occurs when multiple interviewers with competing agendas are involved. Not only does such a situation tend to indicate some degree of an internal power struggle, but it can often make it difficult to get a real answer about what is truly required to do the job. For instance, you might have one interviewer who really thinks that the team is weak in regard to their database work, while another thinks that they just need more people working on their web application. When this occurs, both concerns are likely to come up during the actual interview, often with little or no warning beforehand.

You have to be careful when this happens, because you don't know what kind of office politics are going on. Specifically, you don't know who is right, and how the whole thing will play out. As a result, you'll have to tread lightly to avoid getting tangled up in the mess. Your goal is for neither party to actively object to you and neither to feel that you've taken a side.

Instead, what you want to do when this occurs is ask more questions. Unless other information presents itself, it's equally likely that either party is right, both are right, or neither is right. Toward that end, you'll want to ask questions that illuminate the issues in question. For instance, with the example I mentioned, you might ask what kind of problems they are having on both the web and database ends. It's entirely possible that you will find yourself agreeing, more or less, with one of the interviewers, but try to avoid taking sides. On the other hand, you may get indications that a larger problem is actually causing both problems. For instance, the company might have made a lot of short-term decisions in the past that are hurting both the database and web teams, and both interviewers are describing different facets of the same problem. If you can figure out what is going on behind the scenes, including a rough idea of the circumstances that led up to the current situation, your answers to other questions will be better as a result.

If you are able to ask intelligent questions about what is going on, the disagreement between the interviewers stops being the focus and you are able to steer the conversation toward something more advantageous. In the case described earlier, you now have enough information to ask good questions about the development environment, how decisions are made, and what practices are in place.

Summary

In this chapter, we discussed how to get through the actual interview, including the implicit personality test that you will face while being interviewed. We also discussed some questions that an interviewer needs to answer for themselves before the interview is over. After that, we talked through some body language hacks that you should use to project more confidence. Next, we discussed how to work your way through a whiteboard problem while distinguishing yourself from other candidates. Finally, we discussed a few things that can make an interview more difficult, and how to react to them so that you stand out from the crowd.

The interview itself is the culmination of a long process of preparation, but it is the most crucial part to get right. After all, the people that hire you don't necessarily understand all the work you did to prepare, but they see the results in how well you interview.

What to Learn During the Interview

The goal of an interview is not to convince someone to hire you. It's for the interviewer to determine whether to hire you and for you to determine whether you want to be hired. Both sides are equally important. The interviewer should, theoretically, have their end of things under control. However, you have to make a decision about whether you want to take the job. Not only should you not take a job that isn't a good fit but you need to have some reasonable criteria to arrive at your decision.

Most of this stuff is highly subjective. Personally, I don't really care for corporate cubicle-style environments. They make me miserable. However, I have friends who do extremely well there and are miserable in the sort of small business environments where I thrive. Everyone has different preferences, and you'll need to carefully consider yours in light of any jobs where you interview.

In this chapter, we'll discuss some of the things you should be learning from the interviewer during the course of the interview. Not only does learning this stuff help you learn more about the business but also changes the tone of

© William Gant 2019

W. Gant, *Surviving the Whiteboard Interview*,

https://doi.org/10.1007/978-1-4842-5007-5_10

the conversation. Interviews are far better when they become a discussion of whether you and the organization want to work together, instead of you trying to convince them to hire you.

Why the Right Kind of Organization Is So Important

It's not just about whether you can work with a particular organization, but you also need to make sure that the organization is suitable for you. Depending on your current life circumstances and your goals, any given organization might be more or less suitable for you.

An older, larger organization is more likely to be stable and have good pay and benefits. However, the work may be boring compared to other environments, and the opportunities for growth may be fewer. Smaller organizations may be far less stable and have lower pay, but may offer far better opportunities to improve your skills.

There are also cultural differences to consider. Smaller, newer companies may be a little more "relaxed" in regard to things like coding standards, dress code, and process in general. Older, larger companies may be more restrictive in these things. You can find employers that match your own preferences pretty easily, but you do want to make sure the organization fits in with your personality and goals. While considering this, don't forget that by taking a job somewhere, you are inviting that organization into a sizable chunk of your life.

Risk Tolerance

The first major factor in evaluating an organization is the amount of risk that you are taking on by working there. Earlier, we discussed finding out as much as you can about the company and its quality-of-life factors. While research before the fact is great and can warn you away from companies that are having severe issues, it's unlikely to tell you everything that might concern you. It may be impossible to find anything out about smaller companies, while larger organizations may have recently changed.

Toward that end, you need to be asking questions to determine how volatile an environment really is and comparing that to the information you found in your research. Here are some good ones to start with. You should be trying to get answers to these questions during the interview, although you may not have to ask them directly.

- How stable is the organization's revenue? Has that changed in a big way recently?

- How long do employees typically stay at the organization?

- Has the company recently purchased another company (or been purchased itself)?

- Have there been major changes to the management of the organization recently?

- How much has the organization expanded (or contracted) in the last 18 months?

This may sound like a lot of prying (and it is), but to get a good idea of the risk that a particular job presents. We'll get into why these questions require answers in a minute, but the goal here is to balance your salary against the possibility of suddenly finding yourself unemployed. This doesn't mean that you don't take a job that's risky—some of the best ones I've had have been very risky. Rather, it means that you evaluate the risk and determine whether you can take it.

For instance, when I was single and during the early years of my marriage (before we decided to start a family), I worked at a few small businesses and startups with a lot of volatility. When we decided to have a child, I transitioned to a more stable, corporate job and stayed there for the first couple of years. Afterward, with finances a little more stable and childcare a little less expensive, I started consulting again. A few years later, we started wanting to move a little further out of town, and I started moving toward more stable employment. As I've gone forward and built out the podcast and other interests, it's been important to me to keep the day job as stable as possible, simply so that other things will remain that way.

Over the course of your career and life, you'll probably experience periods of relatively higher risk tolerance and periods where you really don't want the risk. That's completely normal. There's also nothing wrong with having low risk tolerance, as long as you remain happy. There's (probably) nothing wrong with a high risk tolerance all the time, so long as you have a backup plan.

The questions listed previously are designed to help you determine how risky a potential job is. Things like major changes to revenue, recent acquisitions (or sales), and large management changes could foreshadow structural changes to the organization. Sometimes these changes will be very good for you, while other changes could mean that you are the first one laid off, due to being the newest employee. Risk factors often come with at least some chance of a reward, but you have to evaluate which you think is more likely.

Rapid expansion or contraction of an organization can mean lots of problems. It almost certainly portends major changes in company culture over time (for good or ill). It also poses a risk to you, since these things tend to happen more than once. If the company had major layoffs, there is a good chance that more are coming. Similarly, a round of sudden growth tends to create a bit of organizational chaos until things settle down.

It's also good to find out how long a typical employee will stay with an organization. If it is a short period of time, like a couple of years, you may find that there are good reasons for that. In particular, it can mean that the company doesn't do a very good job with raises and promotions. Companies cannot retain good people if they don't pay well.

You might think that this means that companies with very good retention pay well and have a good environment. That might be true, but there is another possibility. You'll often see companies where many employees have been there for 10 years or more, while there is a lot of churn in the newer employees. This often occurs when the long-time employees aren't particularly good, but are comfortable. The newer employees stay until they give up and then they leave. This is often referred to as the "Dead Sea Effect." An organization with these characteristics tends to be highly siloed and resistant to using newer technology. While less of a risk in the short term, they can be a huge risk to your career over the long term, as these jobs tend to go away after you're already well behind with your skills. As a result, you may find yourself out of work and having to improve your skills before you can get another job.

Structure

You should also learn a bit about the structure of the organization. Organizational structure may have deep hierarchies or be relatively flat. While a deep hierarchy typically reflects a larger organization, some smaller organizations have a surprising number of people between the average developer and the CEO. The way that teams are managed has a huge impact on developer quality of life and on the ability of developers to move up in the organization.

Additionally, the depth of the management hierarchy also makes it harder to know what is really going on in the company. If you work at a company without a huge hierarchy, you'll often be able to see the warning signs of layoffs well in advance of them occurring, and you are more likely to have regular contact with the people making decisions. This can give you a lot more opportunities to provide value within the organization, but it also exposes you to corporate politics. Depending on your personality, this is either a bug or a feature…

Things get even more complex if your management structure isn't in the same building (or even the same geographic area). While this can mean that you are largely left alone to do your work, it also reduces your chances of advancement. You'll be spending more time showing your management that you are actually working, and it is easy to end up feeling left out. On the other hand, if your management is remote and you are able to show that you are productive, you'll find you have a lot more freedom at work.

Pay Scale and Compensation

The interview, or shortly thereafter, is when you will find out about your actual compensation. Note that this may be wildly different than what was in the actual job listing. If you do particularly well in a corporate interview, it's entirely possible that they will offer more money to bring you on board.

It goes without saying that you should know what a job pays before even considering taking it. To be honest, you're probably better off knowing whether the pay rate is sufficient before even interviewing. You don't want to waste your time interviewing with an organization that won't pay you enough, unless you think you can get them to make you a better offer.

In addition to the starting compensation, you need to find out about things like benefits, time off, and insurance. While this varies a lot depending on where you live, these things are part of your compensation for working (at least in the United States—insurance tends to be very different elsewhere). In addition, if you are moving, or are incurring other costs for working somewhere, such as the cost of your commute, you have to be rather careful when evaluating a job based on compensation.

Rather than simply having your salary increase, what you really want is for your available disposable income to increase. That is, your spending power after your expenses should be steadily increasing over time, barring any lifestyle changes. We're going to get into a little math here, but we're going to try and keep it reasonably simple. First, an equation for disposable income (not available disposable income, which we'll get to in a minute):

Disposable Income = Total Personal Income – Personal
Current Taxes

The formula is simple and shows the first thing that you need to consider when looking at a job. Your disposable income is the money you actually get to keep after taxes, fees, and the like that you are forced to pay, simply to live and make a certain amount of money. You need to examine how a new job will impact your disposable income, especially if the new job will force you to move, because it's easily possible to get a higher paying job and have less available money. For instance, if you are making $50,000 a year at a 10% effective income tax rate (it'll be worse, but round numbers make math easier), then your disposable income would be $45,000. If you leave that job for a $60,000 a year job with a 25% effective income tax rate, then you'd still have $45,000 in disposable income. While that sounds like it would be the same, the latter would actually leave you worse off, because housing, food, childcare, education, and transportation expenses would probably be higher in the new location. If a new job leaves you with an equal or lower amount of disposable income, you should think long and hard about whether you want to take it. There are reasons you might, which we will discuss later in this chapter, but a decrease in disposable income might make life very uncomfortable if you are unprepared.

However, total disposable income isn't a good measurement of your quality of life. You will still have other, nearly unavoidable expenses that will take away from your disposable income. While an exhaustive list of these would be excessive, pointless, and rather painful to think about, we can at least discuss some of the big ones. To be able to live, you'll have expenses such as the following:

- Food
- Housing
- Transportation
- Medical expenses
- Utility costs
- Childcare (if you have children)
- Insurance
- Education and student loan costs

Many formulas for this sort of thing will try to calculate your discretionary income by using a complex formula using the Federal Poverty Line (FPL) formulas based on the size of your family, your adjusted gross income, and a bunch of other stuff that seems really relevant. We're going to simplify this a lot. Discretionary income calculations can tell you that you have more money when you actually feel like you have less. In particular, things like housing costs, transportation, education, childcare, and medical costs can factor into this to make you feel like you are broke when the equation says otherwise. Instead, we'll consider the following:

Available Disposable Income = Disposable Income − Mandatory Living Expenses

Now, there is probably an economist somewhere who is going to read this and get extremely annoyed by it. But realistically, this is how you, I, and just about everyone we know really look at income. We look at what money is available after required expenses to build some kind of decent life for ourselves. The category of "mandatory" expenses is really subjective. While you might change your definition of "mandatory" over time, you will certainly feel it if your "mandatory" expenses increase. This happens regardless of whether an economist thinks your assessment of "mandatory" is valid or not.

Additional available disposable income gives you options and helps you absorb risk. It makes it possible to save, go on actual vacations, and just generally enjoy your life. This is the number you are aiming to improve, regardless of the other stuff.

When calculating this number, you'll want to include all the expenses required to keep your quality of life at a tolerable level. For some people, this means that you live in a small house in a "safe enough" neighborhood, can afford a bus to work, and maybe go fishing on the weekends. I have friends doing this, and it is easier for them to have a lot of available disposable income. On the other hand, you might really feel that you need to live in a really nice neighborhood, with good schools, and you want to be able to pay for childcare and a vehicle for travel. We're not injecting any judgments of your life choices here—we're just doing the math.

It's entirely possible to move to a situation where your disposable income and your official calculation of discretionary income both increase, while your actual available disposable income decreases sharply. For instance, let's say that you live in a smaller, third-tier US city where the average house payment for a 1500 square foot house is $1000 monthly (yeah, we rounded down considerably) and your new job requires that you move to one of the "cool cities" where the monthly payment for a 1200 square foot home is $3500 monthly, on top of additional taxes, including income tax. Furthermore, that $1000 a month house was in a safe neighborhood, while the $3500 a month house is in a rough neighborhood. This may mean that you now consider it "mandatory" to send your child to a private school for the low price of a cool $24,000 a year. Let's assume that all the other taxes and fees come up to a total of $12,000 a year. The question you need to ask is not "how much did my income go up?" but rather "how much more money do I need to have a similar or better amount of available disposable income for the same quality of life?" For the drastically simplified example earlier, let's say that you are bringing home $5000 of disposable income a month. Your available disposable income would be defined by the following formula:

Disposable Income ($5,000) – Housing ($1,000) = $4,000

When you really calculate this, you'll need to include all of the mandatory expenses on your list, so the real number would be less than $4,000 a month. However, we're simplifying this to help make this less daunting. Now, let's look at the same amount for the "new job."

Disposable Income (x) – Housing ($3,500) – Taxes ($1,000) – School ($2,000) = Available Disposable Income

If we were aiming for the same amount of available disposable income in the new job, we need to do a bit of simple math:

$$\$4,000 = x - \text{Housing } (\$3500) - \text{Taxes } (\$1,000) - \text{School } (\$2,000)$$

Or:

$$x = \$4,000 + \$3,500 + \$1,000 + \$2,000 = \$10,500$$

The same available disposable income requires a pay raise of $5,500 a month, or $66,000 a year. For a family, that's pretty steep. For the purposes of a family, this seems like an indictment of expensive, first-tier, "cool" cities. And it probably is.

However, there is another angle here. What if you are single, childless, and perfectly fine with a long commute from the suburbs of the "cool" city? You are competing with people that require an extra $66,000 a year, simply to be in the same situation they were in before. If you were to take a $50,000 increase, you could outcompete those people on price, save up some significant assets, and then pay cash for a house in the third-tier city later when you decide to settle down and start a family. Without the mortgage, the third-tier city is also far more affordable, so it's entirely possible that living in the "cool" city for a while will help you more in the long term.

Hopefully you can see how playing with the numbers in terms of compensation and expenses can give you some evidence of whether a job is financially better for you personally. It also (probably) makes it obvious that you have more options the fewer "mandatory" expenses that you have. However, you have to balance that with your quality of life, or you may find that you can be miserable, even if you have relative wealth.

These considerations are helpful, but they don't completely cover everything you might want to consider. In addition to the available disposable income a particular position provides, you also need to weigh this against the total hours required for your work. Let's start with some equations to make that understanding a little more obvious. First, we'll start with a simple calculation for your available hourly pay rate (assuming that you are a salaried employee). We're basing this off of what you actually have available after "mandatory" expenses, because that's really a better metric for how much money you are making.

$$\text{Hourly Available Pay Rate} = (\text{Available Disposable Income} / \text{Total Hours Worked in a Year})$$

For instance, if you have $25,000 in available disposable income after your mandatory expenses, you are expected to work 40 hours a week and you get 2 weeks of vacation time in a year, then you could say your available hourly pay rate is $12.50. Every hour you work, you get $12.50 of money to do whatever you want. That's actually pretty good, until you start realizing what it takes in most places to get $25,000 in available disposable income. It's doable if you are frugal and negotiate well, but it is going to be difficult until you have a really good job and low living expenses.

Like the previous equations, it's also possible to have an increase in available disposable income, yet have a substantially worse life. For instance, a $1000 increase in your yearly available disposable income may not be worth it to you when the new job requires 80 hours a week of work. In that case (assuming you went from 40 hours a week to 80 hours a week), your hourly available pay went from $12.50 to $6.50. Looking at just raw numbers, you are better off financially, but the strain that an 80-hour-a-week job will put on your health, your sanity, your relationships, and just your available time is immeasurably worse. It'll be worse than that if the longer hours cause you to have a major health problem, cost you a lot of extra money by increasing expenses (or causing a divorce), or simply make you miserable.

The preceding equation is useful, but there are a few issues with it. First of all, Total Hours Worked doesn't include the time you spend commuting, getting ready for work, and traveling to and from lunch (if you need to get out of the office to remain sane). It also doesn't include the time and expense caused by being away from your residence, which can cost you money and time for everything from childcare, to difficulty in arranging doctor's appointments, to having packages stolen off your front porch. The preceding formula is only a way to attempt to quantify whether one job's compensation is better than another, and it's really only a starting point for things you need to consider. There are a lot of things that are hard to quantify with money, so this is only a partial solution to the problem. This is roughly my thought process, but it doesn't fit everyone.

Advancement Opportunities

In addition to your pay scale and other compensation, you should also consider a position's potential future ability to create income. That income may arrive in various manners and is highly subject to your creativity. You shouldn't just evaluate jobs in terms of what they pay for however long they last. If you only factor pay into your calculations, it's very easy to end up in a job that doesn't help you over the long term.

For instance, if your job pays tens of thousands of dollars more over the course of a year than any other job you can expect to get, yet your job relies on technology that is becoming obsolete, you may find it impossible to get another job at the same pay when that one ends. This can be especially nasty if the job loss is sudden.

It can also be pretty miserable if the job itself doesn't take your career in the direction that you want it to go. For instance, I personally could make more money, especially over the long term, writing software for various situations in finance. I enjoy some of the kinds of work that such a job entails, but I would be miserable with the hours and stress that comes with it. I suspect the same would be true of game development and a lot of heavy corporate programming in general. Ultimately, I would like to own my own software business and be my own boss. A lot of jobs won't help me develop the skills I need for such a venture, and many of them may make it more difficult to get out on my own. The concept of "golden handcuffs" in a job that you hate is a very real risk, and it's one you need to avoid.

Therefore, it's necessary to weigh the possibility of future opportunities against the pay that a job offers currently. Some of this calculation is still pay related. For instance, if a job helps you learn a skill that pays well, it may be worth taking instead of a job that pays more, simply because you can make more later and recover the difference.

The nonfinancial aspects also need to be considered. You (hopefully) have some idea of where you'd like your career to be in the next decade or so. If not, you need to stop right now and figure that out, simply to improve the quality of your decisions. However, most people that have gotten this far probably have an idea of what they want out of life over the long term (even if that may change later).

While the way that you evaluate the longer-term career options that a job represents is almost entirely subjective, there are a few considerations that might help you reach a decision:

- How soon would you like to reach your ultimate goal? You will have to be much more aggressive in pursuing opportunities if the goal is shorter term, while even small improvements can help with a goal pursued over a much longer period of time.

- What skills and mindsets does this position allow you to develop and what are those worth to you? Are these opportunities presented in a real-world environment (as opposed to a classroom)?

- What industry contacts does this position help you to attain? Will you be able to leverage those later?

- What opportunities does the position allow you for self-promotion? Can you build credibility with people where you will need it later?

- Could you develop a strategic partnership with this employer later if you need it?

- Will this opportunity allow you to develop business knowledge that you can use in later jobs?

- Will you have the opportunity to learn management, sales, project management, or other skills that are useful across a broad range of other opportunities in the future?

- Does this position allow you to work remotely or give you more free time to develop your own products or services if you desire to do so?

- Does this position allow you to purchase real estate or other property in an area that you have identified as being a good investment? While this is out of the scope of this book, good real estate investments can make a huge difference in your income if you can get them.

- What opportunities does this position offer you as far as finding a mentor who will help you grow?

- Does this opportunity allow you to learn technology or technological principles that will help you get positions with even more opportunities?

There are a lot of things that can make a job more valuable than simply the money involved. As long as you can pay your bills with the money you make from a job, these nonfinancial prospects should be a huge part of the criteria you consider. You can almost always find a job that pays a little more than your current one, but if you want huge increases in your income, quality of life, and ability to control your career, there are far more important things to consider instead of money.

Training Opportunities

While the previous section focused on the longer-term benefits that a position offers, one especially critical item to consider over the short term is the availability of training in a position. In the fast-moving world of software development, training is critical if you want to continue to be employable. It's also really hard to train on your own off the clock, especially if you have other obligations.

Therefore, it's entirely possible to end up trapped in a particular job, simply because you didn't get any training to help you stay current. Since life is easier with options, even if you don't take them, you'll need to make sure that the job provides adequate training opportunities and that those opportunities will actually allow you to advance. Here are some reasonable questions to ask:

- Does the organization pay for training web sites like Pluralsight, LinkedIn Learning, or Lynda.com? At a minimum, most organizations should do this, simply because it allows employees to learn things as they need them.

- Does the organization pay for learning on the clock? If you are having to study and learn on your own time, your total number of work hours in a year is much higher.

- Does the organization pay for onsite training or for sending you to training offsite?

- What about conferences? Do they pay? Do you have to take conference time out of your time off?

- What about further education? Can you get a master's degree if it would be useful?

- Are there people in the organization who are further along in their career path who would be willing to act as mentors?

Employers offer varying levels of training, and it's important to have an idea of what those are before signing on. You should ask about this during an interview. After all, if you stay there 5 years and don't get any training, you may have to stay there from now on. While you shouldn't necessarily rule out working somewhere based solely on training, you do need to factor training in when comparing potential employers.

Red Tape and Bureaucracy

Another item that you should consider when evaluating a potential workplace is the amount of paperwork, red tape, and bureaucracy that you'll have to deal with as part of your day. While there are lots of people that don't mind a little bureaucracy, those who hate it tend to really hate it. It's pretty rare to find anyone who truly enjoys convoluted, annoying processes at work, unless they either wrote the process or they plan to rewrite the process. If you are neither of these and bureaucracy is something you loathe, you should find out how bad it is before you take the job. A few questions can help you find out how much pencil pushing you will be doing:

- What software development lifecycle methodology do you use? While agile is all the rage, if they answer with "agile," you should ask more questions. Poorly implemented agile is at least as bad as "waterfall" development and can often result in a lot more useless work.

- How do you handle bug tracking? They should already have a process for this. If they don't, you're going to be spending a lot of time either implementing a process or paying the price of not having one.

- How do you come up with and design new features? Many companies have an extensive design process in which the developers are expected to participate. Sometimes these processes are extremely painful to deal with.

- What is your software's lifecycle in the world? If they are continually deploying to the cloud, their documentation requirements and instructions will look very different than a company that deploys to client systems every 6 months.

- How much project management overhead are developers burdened with? Organizations may vary from simply requiring developers to periodically update their status, all the way up to developers spending a ton of time in a project management tool.

- How quickly can a new developer add code and add value to an application? If the answer is "weeks" or "months," they probably have a bit of a process problem that will manifest as paperwork.

- How uptight are they about their time system? A general-purpose time system can really help the organization, but an overly complex and Byzantine time system can quickly destroy developer morale, especially if combined with management that doesn't see it as a problem.

Red tape and bureaucracy is not a reason to avoid taking a job. However, combined with other problems, it is certainly a reason to reconsider. Besides how it wastes your time, you may also be expected to be as productive as a developer who isn't similarly burdened. This can mean excessive working hours or the possibility of a stagnant pay rate as you fail to meet ridiculous requirements.

On the other hand, organizations with no overhead have their own problems. They may have such poor processes that you are constantly trying to figure out what you should be doing, rather than actually being productive. You will need to make judgment call in regard to whether you can handle the bureaucratic workload that comes with any position.

While you may not be bothered by an excessive amount of paperwork, it also has downstream effects on the rest of your team. In many cases, it damages morale in a way that you will notice and that will decrease your quality of life.

Technical Debt

Finally, when evaluating a potential position, you should also try to get an idea of the organization's level of technical debt. While definitions of technical debt abound, what you are really trying to discover is how much of the codebase is crusty, hard-to-maintain, mind-numbing junk. Your definition of that will be different than everyone else's. That's OK, because we're using it to make a subjective judgment about a potential job.

Experience will teach you that most organizations who have any level of technical debt also drastically underestimate the cost of this debt. Technical debt can slow the development of new features to a crawl, trap a team in an out-of-date development framework, and even leave them helpless in the face of new security issues. It's more than just developers disliking old code—technical debt is a time bomb waiting to explode under an organization.

Technical debt also makes it more difficult to hire and retain developers (unless the company wants to pay more). Developers often find it extremely frustrating and would generally rather avoid it. Worse still, technical debt also decreases retention of more skilled developers, meaning that over time the organization will fill up with people who are incapable of fixing the technical debt and incapable of getting away from it. This phenomenon is referred to as the Dead Sea Effect. As a result of this increased rate of turnover, many companies with significant technical debt will be interviewing new developers far more often than their size might suggest.

Technical debt probably bothers you as a developer. While some developers are less concerned by it, every developer has a threshold of tolerance for it. Past that point, the job becomes miserable and you'd rather be somewhere else. It's often not the technical debt itself that directly causes the problem— it's often the organization's reaction to technical debt issues that causes unpleasantness.

Since there is some level at which technical debt causes too many problems to tolerate, you need to have some way of determining how much technical debt an organization has. This can be tricky, since organizations with severe issues in this area are usually in denial about how bad their problem is. Here are some questions that might help you get started:

- How old is your codebase? A really old codebase is going to have lots of little wrinkles hidden in it, no matter how diligently the developers have tried to maintain it.

- How common is it for developers to spend time refactoring or cleaning up code? If this never happens, or if the word "refactoring" seems to trigger something negative in management, their issues are probably severe. Worse, they may still have the management issues that caused the technical debt in the first place.

- How do they prioritize bugs, feature requests from existing clients, code maintenance, and features for prospective clients? This will tell you a lot about how much technical debt is driven by management priorities.

- How would you compare your development team a year ago to your development team now? If the quality of the team is roughly static or on the decline, that usually speaks to the sort of problems with retention that form most of the feedback loop that creates technical debt. Don't expect an honest answer to this question with the team in the room.

- What are your backward compatibility constraints? Either extreme can be annoying here. If they have to support particularly old browsers and operating systems, then they aren't as likely to take risks with stability. If, on the other hand, they don't care about it at all, you'll probably still encounter many of the same problems that occur in organizations with excessive technical debt, namely that problems in the code tend to overshadow your efforts to fix things.

- Has your organization always been comfortable with code cleanup and refactoring practices? If not, how recently did you change your mind? If a company has just recently started cleaning up their code, it's often an excellent time to get hired there. Not only will this help force you to improve your skills, but this also reflects management changes that empower developers. The latter is not only important for your job satisfaction but also indicates management commitment to stop doing some of the things that lead to technical debt.

Technical debt can make a job more difficult than it has to be. Many organizations struggle with years or even decades worth of "temporary decisions written in stone" that have to be managed to varying degrees in order to be productive. However, cleaning up technical debt can be a lot of fun and can greatly improve your code if done in the right environment. You'll only find out whether the environment is right by asking a lot of good questions.

Summary

In this chapter, we discussed some of the things you need to learn during the interview process. When interviewing, you should always be of the mindset that not only do you have to convince them that you'd be good but that they have to do the same for you. Toward this end, we discussed how organizational structures and tolerance of risk intersect with your own life goals. Next, we discussed how pay scale, advancement, and training opportunities are critical to your longer-term goals and how to find out about them. Finally, we talked about how red tape, bureaucracy, and technical debt can be motivation killers and how to find out how bad it is before taking on a job. In the next chapter, we'll discuss some interview red flags that could indicate that a job is unsuitable.

Interview Questions

In addition to the information you should be gathering about the job in the interview, you also need to be prepared for some of the kind of questions that an interviewer will ask. If you followed the previous advice about how to research the company and how to appropriately construct your resume, you will actually end up with a surprising level of control over the questions being asked. Good research informs the structure of the resume you send to a potential employer. A good resume changes the questions that you will be asked so that they work to your advantage.

However, let's be realistic. The first few times you go through the process of interviewing for a job, you're probably not going to prepare adequately. While it isn't optimal, it is entirely normal. Even people that write books on how to get through software interviews were inadequately prepared for their first few interviews (trust me on this one). If you find yourself in this position and reading this chapter, don't worry. You'll still get a lot of information that will help you answer interviewer questions, but you won't have as much control over what those questions are.

© William Gant 2019
W. Gant, *Surviving the Whiteboard Interview*,
https://doi.org/10.1007/978-1-4842-5007-5_11

Here Be Dragons

First off, it's time to talk about downsides. You may well have purchased this book out of a sense of fear of the sort of probing questions that are commonly asked in interviews. If you combine that with a lack of experience in the industry, the prospect of an interview is extremely daunting. It's like the old ocean maps, where certain areas were thought to be dangerous and were labeled with "Here be dragons" to warn sailors not to go there.

There's also no way to overstate this. No matter how well prepared you are, if you bomb the interview badly enough, you won't get the job. It also doesn't matter how long you do this, eventually you will fail miserably at an interview. It happens to everyone and is a critical part of the learning process. If you understand this going in, it not only makes the failure easier to deal with, but it makes recovery from failure a much quicker process.

There are a lot of things that can go wrong in an interview. Sometimes nervousness sneaks up on even the most confident people. Sometimes you'll say something offensive that you don't intend, end up an hour late because of traffic, or they just won't like you for whatever reason. Most of the things that can go wrong in any other human interaction can also go wrong in an interview.

However, with all the potential downsides of an interview, there are a lot of upsides. A good job can change your life for the better in a way that enhances a lot more than just the amount of money you have. Life is a lot easier when you aren't broke, and improvements in your job situation can make you feel better about life in general.

Because of both the high risks and potentially high rewards, you really want to do as much as possible to reduce the amount of anxiety you feel going in. You are far more likely to have a negative interview experience if you are overly worried about the outcome. On the other hand, if you aren't as worried, it's a lot easier to avoid doing something that you need to worry about in the first place. There are a few things you can do that will make it easier to relax in any particular interview:

- The most straightforward thing to do is "simply" to make sure that you have a lot of options. While it's anything but simple to actually get to the point that you have more options than you can even consider, having even just a few other options will help you be more relaxed. You should never be interviewing with only one or two companies—the more options you have, the less power any one option will hold over you.

- Be overprepared. Dress well, arrive early, and research the company thoroughly. You'll feel less anxiety about making a mistake. People tend to see what they expect, and people tend to expect what their first impression led them to expect.

- Have your own questions to ask. This changes the conversation so that it's more about finding a situation that works for both parties. This is far better than setting up the conversation to only be a discussion of your suitability for the job.

- It's also important to understand that any potential job is not the only one that you'll ever have. They are all stepping stones to the next thing. As a result, there isn't such a thing as "the only good job on the market." There are a lot of jobs that are good enough to get you to whatever goal you have. Some may be more or less ideal, but don't allow yourself to feel a sense of scarcity about any of them.

Worrying excessively about the potential for an interview to go badly can actually cause it to go badly. However, if you can manage to be a little more outcome independent, you'll find interviewing to be far less uncomfortable. Don't worry too much if you can't overcome your anxiety early on—you will become more relaxed over time, especially if you prepare well.

Unpleasant Interview Tactics

In addition to the difficulties posed by your own nervousness, there are certain things that happen in interviews that you might want to be prepared for. While uncommon, they can really ruin your day. These tactics can indicate that a job is unsuitable and that the company is not a good place to work. While researching this book, I spent a lot of time hanging out in forums dedicated to hiring developers. In that time, I saw a lot of nasty tricks that interviewers will pull during interviews to help them disqualify candidates. While some techniques were actually reasonable and sane, I did see some stuff that I found inappropriate. In this section, I want to talk through some of the stuff I saw. It was a real eye-opener for me, and it might be for you.

All these tactics are intended to filter out people that might be a problem for the organization. These problems can range from having a bad temper, to being the sort of person who causes human resources "incidents," to people who might go to the police if they see something that doesn't look right. While these are reasonable worries to varying degrees, the problem is with the tactics, in particular their ability to make an interview extremely unpleasant.

These tactics can also trip up reasonable, sane people who are applying for a job, simply because they are caught off guard, are trying to be polite, or are trying not to offend the interviewer. The following are some terribly misguided strategies that interviewers may use in an effort to evaluate your suitability for a position:

- They try making mildly bigoted remarks (or remarks that might be construed that way) to see if the interviewee will play along. While nobody wants to hire a bigot (or worse, work for one), some interviewers will do this. If you see this, it's probably better to try and find a job elsewhere, as it indicates more problems than just being terrible at interviewing.

- They might try mentioning "sketchy" things like fraud, drug deals, and the like to gauge your reaction to them. This may be a method of assessing your honesty, or it may be a method of determining how willing you might be to do unethical things on the job. Be very careful if this happens and try to find out which it is. If this occurs, you may even want to consider reporting it, depending on the subject matter. At the least, this can be a good reason to look elsewhere for work.

- It's normal for an interviewer to ask you about your previous employer, but some interviewers will pry far more deeply on that. This tactic can be used for everything from trying to get you to say something bad about your previous (or current) employer to facilitating industrial espionage. Later in this chapter, we'll be discussing inappropriate interview questions like this one.

- They might tell you that they don't think you'll be a fit for a development job and then offer you another job (for instance, in support), and then decide that you aren't really motivated enough if you express interest in the other job. This is a particularly slimy technique that takes advantage of people who are desperate.

- You may also find that some interviewers don't really answer the questions you ask during an interview. This evasiveness can mean a number of things. It might simply mean that the interviewer doesn't know the answer, but it could also show that they are intentionally trying to deceive you.

There are a lot of things that can happen in an interview that might indicate that the environment in a particular organization is going to be unpleasant. If you experience any of these things during an interview, you probably don't want to pursue that position any further. I've seen all of the items previously listed in interviews at one time or another and have heard horror stories from the people who took those jobs. Even if you are desperate, these are almost never worth the trouble.

Technical Questions You'll Be Asked

You'll also be asked technical questions when you are being interviewed, with the intention of evaluating your skill level. While not all of them will do a good job of showing your real skills, they are still an ever-present part of any software development interview. You must prepare for them. In this section, we'll discuss some types of questions that are commonly encountered.

First of all, you'll be asked some computer science questions. These are general-purpose questions used to weed out unqualified candidates. They also have a tendency to be about topics that are largely irrelevant for your day-to-day work. Here are some examples:

- *Which search algorithm is most efficient, the one that is O(n!), O(logn), or O(n)?* You'll get asked this question and almost never use this stuff in a real job (in 20+ years, I've needed to break down the big O notation for something exactly twice, outside of interviews). Basically, the gist of it is, as "n" increases, which algorithm produces the smallest number? In this case, it would be O(logn), since the result of it increases logarithmically with respect to n. This is followed by O(n), which represents a linear growth rate with respect to n. The worst is O(n!), which represents a growth rate of n factorial. Your best bet for this one is just to memorize and go on with life.

- *What are the basic concepts of OOP (object oriented programming)?* These would be encapsulation, abstraction, inheritance, and polymorphism. While explaining this thoroughly is outside of the scope of this book, a quick Internet search will point you toward reference material that will help.

- *What is the difference between overloading a method and overriding it?* If you overload a method, then multiple versions of it are available, including the original. If you override a method, then only your override is present (barring other overloads).

- *What is a stream?* A stream is a sequence of data.

In addition to basic computer science questions, you'll also be asked questions about the programming stack that the company is using. This will be everything from their front-end technology to the database technology in use and may even include things like libraries and source control systems that the company uses. Because these can vary considerably across both platforms and time, answering these questions is outside the scope of this book. Instead, if you've done the organizational research suggested in the previous chapters, you'll already have a rough idea of what technology they are using. From there, you should go to a search engine and look for common interview questions for all the various tools and languages that the organization is using.

You will also be asked about development methodology. This can be anything from how you use source control to questions about agile development processes. Like the questions about the software stack listed earlier, you'll need to use a search engine to find appropriate answers to these questions. However, the development methodology that a company uses is often much more difficult to determine before the fact. You are far better off preempting this discussion by asking your own questions.

Nontechnical Questions You'll Be Asked

You're also likely to be asked a lot of nontechnical questions to determine how well you will work with the team. While many of these questions don't do a particularly good job of finding out about team performance, people tend to ask them anyway. You should practice your answers to these questions so that you are prepared for them. The key to effectively answering these questions is honestly reframing them in such a way that the answer works to your advantage.

Here are some samples of the kind of questions you might be asked. Along with the questions, I've added an example of a bad answer and a good answer. We'll discuss the reasoning behind the answers afterward.

1. *What would you say is your greatest strength?*

    Bad Answer: My ability to quickly write code.

    Good Answer: My ability to work with clients to determine their needs and then execute on those needs.

2. *What would you say is your greatest weakness?*

    Bad Answer: I get really angry when I find bad code.

    Good Answer: I'm very passionate about code quality and get upset with myself when I don't measure up to my own standards.

3. *How do you resolve issues with a team?*

   Bad Answer: I try to figure out who screwed up, fix it for them, and then teach them what happened.

   Good Answer: I look at it as an opportunity. Team problems come from process problems, and if you can help fix the process, everything improves.

4. *How do you troubleshoot issues with other people's code?*

   Bad Answer: I look for places with bad coding standards and rework the code to match standards.

   Good Answer: I run the code and then look at logs and debugging information to start trying to figure out what is causing the error.

5. *What have you done that you are proud of?*

   Bad Answer: This one time in college I... (assuming college was more than a few years ago) I wrote a paper in history class that the teacher thought was really clever.

   Good Answer: I was able to save my previous employer $X by changing the way we were using caching on our web servers.

6. *What would your perfect job look like?*

   Bad Answer: Something that isn't true.

   Good Answer: Something that is true.

7. *What is your current salary?*

   Bad Answer: An answer.

   Good Answer: I think this job is a completely different set of circumstances than my current one. There are lots of differences in insurance, the time I'll spend commuting, and the stress level of the position. What would be the typical pay rate for this position in this organization?

8. *Why did you leave your last job?*

   Bad Answer: Because upper management was stupid and were making a lot of bad decisions that were bad for the codebase.

   Good Answer: Because things changed at the company to the point that it no longer felt like a good fit. Some decisions that were made caused me to believe that the job was going to be significantly worse and more risky in the near future.

There are a few things at work here. First, all these questions are terrible questions, but are still good examples of the kinds of things you'll be asked. If you read over the good answers and bad answers, there are a few things that jump out. Let's start with the bad answers. The bad answers have a few things in common. The most obvious issue is that they not only give a bad impression of the interviewee, but that they also don't move the conversation toward a dialog. The worst thing you can do when answering a question is to set it up so that whatever opinion the interviewer forms is the only one that they can form. Instead of doing that, you want to hedge your bets so that you can have a real conversation as an answer. This significantly reduces the risk of an answer being misinterpreted and costing you an opportunity.

If you practice steering interview answers toward creating a dialog between yourself and the interviewer, you will have a much larger margin of error in regard to your answers for questions like the previous.

There is also something similar about the good answers. In these cases, you not only try to drive the conversation toward a dialog but the answers are also designed to be as positive as possible. Even when delivering a truthful answer that is unpleasant, there are ways to phrase things so that you don't make a negative impression. For instance, even though upper management may be making some stupid decisions that make you want to quit your job, you may be better off saying that the environment shifted in such a way that the job became less of a good opportunity.

Finally, there is one question in there that is different than most of the others. When an employer asks what your current salary is, it's almost never a good idea to answer that question directly. While an organization will almost certainly offer you a salary that is 10% higher than your current one, answering this question at all destroys a lot of your negotiation power. Once you state a number, you are often stuck with it. If you instead change the answer to foster dialog, you can actually find out what the going pay rate is and have a discussion about that instead of your current salary.

Inappropriate Questions

Finally, there are a few questions that interviewers really shouldn't be asking in the first place. These questions are hazardous to the interviewer and possibly their organization. In addition to improper questions, some comments that interviewers can make are also inappropriate. It's important to know that these things are out of bounds, but it is not necessary to practice your responses to them. In many cases, these questions are illegal to ask, as they can be used to introduce bias into a hiring process. If you see any of these, you're probably better off just going to the next opportunity. However, do be careful to listen to the entire question if you happen to get asked any of these. For instance, they may ask you if you are from a particular country because your accent reminds them of their parents, or because they know of an upcoming project that will require someone who speaks your native language. Sometimes these things are a bit more subjective, and you may not want to react too quickly.

They may ask about age, gender, gender identity, marital status, religion, ethnicity, country of origin, or if you have kids. Asking if you have kids is probably the biggest one. While in polite conversation, people may ask about this stuff with the intention of just learning more about you (with varying degrees of tact), in an interview, it's a really bad idea to do so. However, you'll probably encounter people that do this, especially if they aren't very good at interviewing in the first place. It's not necessarily a reason to avoid an organization, either, as this sort of thing is especially common in smaller companies who really don't know better (whether they should know better is another issue entirely).

Asking who will take care of your kids while you work is something that you'll see occasionally, especially if you are female and mention your children. This is daft even in polite conversation, but it may indicate that the organization had prior bad experiences that they attributed to working mothers. On the other hand, if the company has daycare available on site, you may have already convinced them that you are a good fit and they may be trying to convince you.

Some particularly insensitive interviewers may ask about medical issues, such as scars, body marks, abnormalities, and disabilities. This is probably well out of bounds unless the job has some special requirements that make it reasonable to ask (for instance, you have to lift heavy things regularly). Those reasons are few and far between for developers.

If you are in America, it's also inappropriate for an organization to ask about how frequently you are deployed for Army Reserve training.

Another fun conversational topic that occurs from time to time is asking the interviewee whether they are planning to have children or not. While this is almost always out of bounds, you'll run across the occasional interviewer who asks anyway.

Usually, companies will also want to know whether you have a criminal background or not. However, they should not be asking whether you've been arrested, but whether you have been convicted. Don't panic if you see this question and you have a conviction. It doesn't necessarily mean that people won't hire you—I've worked with developers who had felony convictions in their past, and they were some of the best and most loyal employees. Other interviewers may well have seen the same thing.

It's also considered inappropriate to ask whether you have citizenship, instead of asking whether you are legally allowed to work, or at least it is most of the time. It's also something that interviewers get wrong, as many of them aren't necessarily aware that you can work without a citizenship.

An interviewer might also ask whether you rent a home or own one. They may also ask about your credit rating, debts, and the like. There's not a lot of reason to do this, except for very sensitive positions with very high security constraints where these factors could constitute risks.

A final thing that may come up would be your use of sick leave, vacation time, and worker's compensation on a previous job. While these things certainly could impact your ability to do a job, there are a lot of limits around what a company can ask about these things, due to the risk of them being abused.

The inappropriate questions discussed in this section are good examples of things that you shouldn't be asked in an interview. However, they might come up. When they do, it's a good idea to have some thoughts as to how you might handle them, but it isn't a good idea to dwell on them too much. Interviewers can do some stupid things sometimes, and these things are worth planning for if you think they might come up in the interview. Even if you aren't asked these questions, if you are prepared to be asked them, it will help you appear more confident.

So, how do you answer these types of questions? This depends a lot on why you think you are being asked. One possible solution would be to simply answer them directly and clearly. This does a few things. First, if the question is obviously inappropriate and discriminatory, it might put you in a better position legally if you are discriminated against. It also shows a degree of openness (rather than defensiveness) that might help in some situations. While these questions are not at all appropriate, sometimes it's better to assume that the interviewer is well intentioned and making a mistake. In many of the cases we previously listed, the interviewer might simply be curious about you as a person, which is a good sign.

On the other hand, if you are already getting the feeling that the interviewer knows good and well what they are doing, you might simply want to politely state that this question doesn't seem relevant and see what happens. The response that you get afterward will make it very clear whether you want to work there or not. Whether you make some kind of complaint afterward is up to you, but bear in mind that doing so is at best distracting. From watching this sort of thing over the years, I can tell you that usually an organization that is run in a discriminatory fashion tends to be run poorly in a host of other ways. While you may be rightfully angry about the inappropriate questions, there are other organizations out there that would be happy to have you. It's better to reach your own goals than to worry about organizations and people who can't contribute to them.

Summary

In this chapter, we discussed the types of questions that an interviewer will ask you during an interview. While it's impossible to predict every possible question or situation that might occur as a result of the interviewer trying to learn more about you, there are some general categorizations that make sense. We started out by discussing the risks of a bad interview, along with how to change the conversation in your favor. Next, we discussed some dirty tricks that interviewers will occasionally play to trip you up and cause you to make a mistake. After that, we discussed the general technical questions that development interviews often involved, followed by the nontechnical interview questions that often make an appearance in a technical interview. We followed that up with a discussion of some questions that are out of bounds and that might indicate that a particular job is not a good choice.

Additional Resources

The previous chapters gave you a lot of information about how to prepare for software development interviews, how to research potential employers, and how to ask the right question during the interview. However, for many of you, that may not quite be enough to get the level of confidence that you need to feel comfortable. In this chapter, I want to talk about some resources that I've found that are very helpful for getting through both the whiteboard interview and interviews in general. Some of these resources also have quite a bit to offer an aspiring developer, even after they've gotten their first job. This is nowhere near an exhaustive list, as there are hundreds of web sites, podcasts, blogs, and YouTube channels that are designed specifically to help developers get started on their careers. However, these are enough to get started.

Podcasts

I would be remiss in failing to mention my own podcast, especially the early episodes, as a resource that you'll find useful on your journey into software development. Myself and BJ Burns (who was my apprentice during the early days of the podcast) started this before he got his first "real" programming job. Each week, we talk through a topic around software development, covering everything from interviewing to soft skills, to security, to algorithms. The goal of this podcast is to help create better developers by having a more junior developer and a more seasoned developer discuss a topic and try to

© William Gant 2019
W. Gant, *Surviving the Whiteboard Interview*,
https://doi.org/10.1007/978-1-4842-5007-5_12

explain it in actual English (versus the overly technical descriptions you get from tech books). As of the writing of this book, we've been going strong for over 3 years. You can find us at www.CompleteDeveloperPodcast.com.

Another podcast I would suggest is "Junior Developer Toolbox." Currently featuring my friends Dave and Erin, the Junior Developer Toolbox is a podcast about building, maintaining, and surviving a career in software engineering. It's about real-life junior developer work in the trenches of software development. You can start checking this out at www.JuniorDeveloperToolbox.com.

Meetups

My friend BJ and I also have a meetup group in Nashville called Developer Launchpad. At this monthly meetup, developers working with a variety of languages and platforms work through practice problems. We have a problem of the month that we put on the Developer Launchpad web site, which is the recommended problem for the meetup. The monthly problem has three levels of difficulty, depending on the skills of the attendee. We also practice working through whiteboard problems if anyone needs to do so. Previous months' problems can be found at www.DeveloperLaunchpad.com. Previous months' exercises are still useful for code katas.

Practice Questions

If you are nervous about whiteboard interviews, you probably are also a bit nervous about the general process of the software development interview. If that's the case, I'd also recommend checking out *Cracking the Coding Interview, 5th ed.* by Gayle Laakmann McDowell.[1] This book goes through a large number of developer interview questions and walks you through the process of solving these problems. If you are doing the code katas as recommended in this book, there are plenty of good options there to use as practice material.

Continuing Education

Once your job search is successful, you'll get a job. While at the moment that may look like the goal of your journey, it's actually just the beginning. If you truly want to be the best you can be, simply being employed is not going to be enough. To be your best, you have to continually be improving yourself. The team at Simple Programmer has created a lot of training material for those seeking to have better, happier careers, along with better, more balanced

---

[1] CareerCup, 2013.

lives. While there is paid stuff here, a lot of the free material is very useful for preparing for an interview and dealing with very specific problems you might encounter in your career.

Not only have John Sonmez and Josh Earl put together a tremendous amount of material in the development space, but they've also helped many other aspiring authors to build their own careers (including myself). The first code I was paid for writing was nearly 20 years ago, and I still get useful, actionable information from the various Simple Programmer channels. The Simple Programmer team not only produces things to help developers be awesome, but they also enable those developers to grow into good mentors for other developers. If putting virality into awesome sounds like your jam, you should start checking them out at www.SimpleProgrammer.com.

I would also highly recommend both of John's books: *The Complete Software Developer's Career Guide*[2] and *Soft Skills: The Software Developer's Life Manual.*[3] Frankly, both of these books should be issued to new developers when they get their first work computer.

Interview Cake

The next logical step after working your way through this book is Interview Cake. The premise of Interview Cake is that the programming interview is a winnable game, and it is one that they teach you how to play well. There is more to an interview than just the whiteboard interview, and Interview Cake teaches you how to get through it.

While this book has been about how to get through the mechanics of the whiteboard interview and how to prepare for the process, Interview Cake actually walks you through a more extensive set of development problems. Not only do they offer programming interview questions by language, but they offer them by company (such as Microsoft, Google, Facebook). If you are planning on getting an interview with any of the companies that are leaders in this industry, Interview Cake can take you from merely ready to crushing the interview. You can take the next step at www.InterviewCake.com.

CodeNewbie

CodeNewbie was started by Saron Yitbarek and features a community of developers, the CodeNewbie Podcast and the Base.CS podcast. They also host weekly Twitter chats that are very helpful for aspiring developers and produce the Codeland conference.

---

[2] John Sonmez, *The Complete Software Developer's Career Guide* (Simple Programmer LLC, 2017).
[3] John Sonmez, *Soft Skills* (Manning Publications, 2014).

- CodeNewbie.org
- CodeNewbie.org/podcast
- https://dev.to/basecspodcast
- http://codelandconf.com/

CodeKata.com

CodeKata is an excellent web site with loads of practice problems that are perfect for using in code katas (hence the name). If you practice these regularly, you *will* do better on real development tests. These problems even include things you won't run into during a typical code camp, so they can also be helpful for closing the knowledge gap between yourself and someone with a computing-related degree. Check them out at www.CodeKata.com.

Project Euler

If you are looking for more difficult practice problems, you should check out Project Euler (https://projecteuler.net/). This web site has some programming problems that are a bit more difficult and require a lot more math. If you are interviewing for jobs that are a little math heavy, then this web site will give you a lot of problems to work with.

Companies That Don't Do Whiteboards

Look, I get it. Whiteboard interviews are annoying enough to deal with that you may want to avoid them entirely. If that sounds like something useful to you, be sure and check out a curated list of companies that don't do whiteboard interviews. There are hundreds of options and at least some of them are bound to sound interesting to you. Check it out at https://github.com/poteto/hiring-without-whiteboards.

If you do go this route, a lot of the material in this book will still be useful to you. At the very least, keep up with your practice with code katas and the like. The lack of whiteboard interviews doesn't necessarily mean that things will be easier—it just means that they are doing something else.

Index

© William Gant 2019
W. Gant, *Surviving the Whiteboard Interview*,
https://doi.org/10.1007/978-1-4842-5007-5

Printed in the United States
By Bookmasters